# Biotech Entrepreneurship, Precision Medicine and Artificial Intelligence (AI)

Simon S. M. Chin

Design by Simon Chin

Manufactured in the USA

2025919046

Library of Congress Cataloging-in-Publication Data

Chin, Simon S. M.

Biotech Entrepreneurship, Precision Medicine and
Artificial Intelligence (AI)

eBook ISBN: 979-8-9924674-6-8
Paperback ISBN: 979-8-9924674-7-5

Hardcover ISBN: 979-8-9924674-5-1

# Table of Contents

# Dedication

*To my beloved mother*

*A mother I love wholeheartedly and respect deeply*

# **Preface**

This book is for a global audience of mostly young adults and adults. In 1999, I founded Iris Biotechnologies to improve breast cancer diagnosis and treatment. Iris started trading as a public company in 2008. In 2014, I founded a precision medicine company, Iris Wellness Labs, to provide in-depth scientific analysis to give insights to patients and doctors facing complex medical conditions using whole human genome (3 billion base pairs,) human microbiome and approximately 900 metabolites.

Iris Biotechnologies applied for patents all over the world, and all our patent applications were granted as patents. I have three patents on "Artificial Intelligence System for Genetic Analysis." In the biotechnology industry, patents can make or break your company. One patent could be worth billions and its application can help millions of people. The highest revenue-generating drug of all time is Humira, an arthritis drug that has generated over $200 billion in lifetime sales.

If you want to learn what you can do to fight against the scourge of pandemics, cancer and other diseases, then this book is for you. The transition into the age of artificial intelligence (AI) is accelerating and we are losing our privacy. AI can be very useful in helping us to improve disease detection and treatment and almost all areas of our lives. However, AI also has the potential to destroy humanity, as we know it. It is clearly an existential threat. We must learn how to discern AI bravely and carefully.

The right people came into my life at the right time to help me achieve many things. University of California, Berkeley was ranked #1 for Chemistry in the country in 1977, and it was the only university to which I applied. I got in and was

given the Alumni Scholarship based on merit. My faculty
advisor, Dr. Glenn T. Seaborg, Nobel Laureate in
Chemistry and Second Chancellor of UC Berkeley, gave
me an opportunity to do research with him at the Lawrence
Berkeley National Laboratory starting in my first quarter as
a freshman. I was very lucky and grateful.

After graduating from UC Berkeley with a degree in
Chemical Engineering, I was hired by the first company
that I interviewed with in Silicon Valley. My trajectory
there was swift – starting as an engineer, I was promoted to
a managerial role overseeing manufacturing within just 9
months, doubling my salary by the end of my first year of
employment. Assuming responsibility for the company's
largest operation 6 months after the first promotion, we
experienced a period of great success and rapid growth.

I simultaneously pursued an MBA at Santa Clara
University while working full-time - juggling
employment's demands with graduate studies' rigors. I
actually enjoyed the synergy of applying what I learned in
school to my work and what I learned at work to my school
projects almost on a daily basis.

My goal is to share my more than twenty-five years of
entrepreneurial experience in biotechnology, precision
medicine and artificial intelligence. I want to give back,
inspire hope and encourage people to work towards a better
future.

I am very grateful for the amazing opportunities I've had in
life. Many people have shared their wisdom and showed
me love and kindness. I am grateful to all my mentors. My
life's journey also had many hardships and injustices that
made me stronger and prepared me for the future. The
United States became the most innovative and successful

nation in human history by attracting some of the best minds from around the world.

I am grateful to Dr. Wing Hsieh, Mr. James Moshofsky, my sisters Grace Osborne and Catherine Cheah, my nephews Joshua and Daniel Osborne, and others for helping me to review different parts of the book and publish this work.

# Introduction

We live in a Dopamine Culture of clickbaits on reels of short videos on TikTok, YouTube, and other platforms, scrolling and sending short texts on iPhone, Android, and other smart phones and tablets. We also swipe on various apps and gamble with ease on our choice of communication devices. This culture is a far cry from reading books and newspapers, writing letters, and having real and lasting interpersonal relationships such as long-term friendships and marriage. Dopamine is pleasurable and addictive. Having dopamine easily all the time is not good for us physically, mentally, or spiritually.

I am struck by the many perilous challenges facing humanity today – challenges like the pandemic, climate change, tensions between countries, global debt crisis and disruptions caused by new technologies such as artificial intelligence. We are all in the same ship and each of these gaping holes must be fixed concurrently. I hope that sharing my experiences and perspectives in this book can uplift and inspire others.

The ability to discern is one of the most important skills you'll ever learn. It involves seeing, recognizing, or apprehending something by sight, another sense, or the intellect. Having the skill to discern will help you to succeed in life. We can learn to discern and then teach our children how to discern, who to trust, who not to trust, and why truth matters. Children are our future and we must help them to succeed.

In our rapid transition to the age of AI, I believe humanity stands at a critical crossroads. The choices we make collectively will shape the fate of our species. This is the last call.

My primary allegiance is to truth –the truth of science, the truth of history, and the truth about our present circumstances. I hope it enriches others' understanding and personal journeys. I consider this book to be the most important work of my life.

My perspective comes from my own lived experiences, and your perspective may differ from mine. I do not intend to judge but simply to share what I have learned in the hopes that it inspires you to re-examine your priorities and life's purpose in light of the precious gift of life itself.

All humans begin their lives at conception. From a biological standpoint, conception at fertilization marks the beginning of a new genetic individual. At this point, a unique genome is formed, setting the stage for potential human development. There is no debate or controversy about that.

Between conception and birth, there are disagreements as to when the growing cells can be called a human. Abortion ends the development of cells that could result in the birth of a baby. There is also no debate or controversy about that.

Whether people argue life began at birth, at 120 days, or at any other time between conception and birth, DNA already defines a person's genetic identity at conception. That is a fact. People can believe whatever they choose to believe, but that won't change the truth. We live in a time of deceit, and we must be extra vigilant, especially with AI used to create "Untruthful clickbaits" online.

As humans, we have limitations. According to our understanding of biology, human life, in our times, is limited to approximately 120 years. Research has shown that when all of the cells ever created in the human body

are multiplied by the average time it takes for cells to reach
the end of their lives, you get roughly 120 years. This is
called the "Hayflick limit, the maximum number of years a
human can expect to live." Leonard Hayflick is a professor
of anatomy at the University of California, San Francisco,
and formerly a professor of medical microbiology at
Stanford University.

Our universe consists of 5% visible matter, 27% dark
matter, and 68% dark energy. Hubble Telescope and James
Watt Space Telescope (JWST) show that our sun is one of
approximately 100-400 billion stars in the Milky Way
galaxy. Our Milky Way is one of 200 billion to 2 trillion
galaxies in the universe.

It is improbable to think that life exists only on Earth. The
human life span, less than 120 years, is just a blink of an
eye in cosmic time. Humans cannot experience the fullness
of life without Earth or something like Earth. That's why
we must do what we can to preserve our world.

# Section 1: Biotech Entrepreneurship

# Chapter 1
# The Seeds of Entrepreneurship

My entrepreneurial journey began at the tender age of twelve on the sidewalks of Burma (Myanmar), where I ran a small book lending business. Customers would borrow novels and comic books for a nominal fee. That modest venture sparked a lifelong passion for innovation one that eventually carried me to the frontiers of biotechnology, artificial intelligence, and healthcare in California's Silicon Valley.

For the past 25 years, I have dedicated myself to developing cutting-edge technologies in the biotech and healthcare sectors, striving to make a positive impact on the world. I've been driven by a deep belief in the power of science and technology to transform lives.

Four years before the completion of the Human Genome Project, I founded Iris Biotechnologies in 1999 with the vision of leveraging cutting-edge genomic technologies and artificial intelligence to transform healthcare. Iris Biotechnologies was the first company to apply AI and gene expression analysis to personalized breast cancer treatment pioneering an approach that helped doctors make more precise, data-driven decisions.

I then founded a wholly owned subsidiary, Iris Wellness Labs, in 2014 with the goal of providing in-depth scientific analysis to offer insights into complex medical conditions like cancer, heart disease, immune deficiency, diabetes, and obesity for patients and doctors at leading health centers. Our analysis based on whole-genome sequencing, microbiome profiling, and metabolite data was far ahead of its time and remains uncommon in most U.S. hospitals

2

today. Yet with the rapid advancement of AI, integrating this depth of personalized data into everyday medical care is becoming increasingly feasible.

Yet, for all the technological breakthroughs and business successes, what I value most about my entrepreneurial journey are the lessons learned and relationships forged along the way. The process of building from the ground up, uniting a team around a shared vision, enduring challenges, and celebrating progress those are the experiences that have shaped me most deeply as a leader and as a person.

Looking back, I can trace a clear thread connecting my various entrepreneurial pursuits, from that first book-lending business in Burma to my current roles in the biotech industry. It is a thread woven of curiosity, creativity, and a deep desire to make a positive impact on the world. These are the qualities that I believe lie at the heart of the entrepreneurial spirit. Successful entrepreneurs tend to share some key qualities and mindsets:

1. Problem-solvers: They have a knack for identifying problems and finding innovative solutions. They think creatively and are not afraid to challenge the status quo.
2. Risk-takers: They are willing to step out of their comfort zone and try new things, even if success is not guaranteed. They view failure as a learning opportunity.
3. Hard workers: They put in the time and effort necessary to turn their vision into reality. They are persistent and do not give up easily in the face of challenges.
4. Leaders: They inspire and motivate others to buy into their vision. They build strong teams and create a culture of collaboration and shared purpose.

5. Lifelong learners: They are curious and always seeking to acquire new knowledge and skills. They learn from their experiences and from the expertise of others.

Significantly, entrepreneurship knows no age, background, or experience level. History is full of examples of successful entrepreneurs who started young (like Bill Gates and Steve Jobs) or launched second acts later in life (like Ray Kroc with McDonald's). What unites them is the courage to act on an idea and the grit to see it through.

If you dream big, embrace challenges, and want to make a difference, then entrepreneurship could be for you. Cultivate problem-solving, calculated risk-taking, hard work, leadership, and continuous learning these qualities will guide you forward.

The entrepreneurial journey demands courage, resilience, and passion. You will face doubt, fear, and exhaustion and also moments of exhilaration, breakthroughs, and pride in creating something entirely new.

As you embark on your own path, remember that every great business began as a simple idea. Your vision, passion, and persistence might ignite the next world-changing innovation. Trust your instincts, but stay humble and open to learning. Seek mentors and advisors who can guide you, yet don't be afraid to blaze your own trail.

In the chapters that follow, we'll explore the unique challenges and opportunities in biotech entrepreneurship: building a strong team; defining market strategy; protecting intellectual property; securing funding; cultivating resilience; maintaining integrity; fostering collaboration; and leaving a lasting legacy.

Before diving into those specifics, take a moment to reflect: What problem drives you? What unique skills or insights do you bring? How will you apply your creativity and determination to make a positive impact?

Remember, entrepreneurship is not just about building a successful venture it's about personal growth, pushing boundaries, and creating value for others. It's a continuous journey of learning and self-discovery.

Whether you're just starting or already on your entrepreneurial path, there's always room to learn, grow, and make a difference. Today's biggest societal challenges healthcare, climate change, education, social justice need innovative solutions and entrepreneurial spirit. Your ideas, passion, and willingness to persevere are the catalysts for progress and positive change.

So embrace and nurture your entrepreneurial spirit. Let it guide you toward creating something meaningful and impactful. The road ahead may be long and winding, but it is a journey worth taking and your entrepreneurial journey could change the world.

# Chapter 2
# The Biotechnology Frontier

When I founded Iris Biotechnologies in 1999, the field of genomics and personalized medicine was just beginning to take off. The Human Genome Project (HGP) a massive international effort to map the entire human genetic code was still underway, with the final results released in 2003. This landmark achievement laid the groundwork for a new era in biotechnology and personalized care.

The HGP revealed that humans have approximately 20,500 genes encoded in about 3 billion base pairs of DNA. This "instruction manual" for building and operating a human being opened up incredible opportunities for understanding disease, developing targeted therapies, and moving toward more proactive, individualized healthcare.

I recognized the potential to integrate genomic data with emerging technologies in data science particularly artificial intelligence and machine learning to transform medicine. What if we could predict disease risk based on a person's unique genetic profile? What if treatments could be tailored precisely to each patient's molecular makeup? These questions formed the core vision behind Iris Biotechnologies.

However, navigating the complex world of biotechnology as a startup came with significant challenges. The science was advancing rapidly, the regulatory environment was rigorous, and bringing new diagnostics and therapies to market was both time-consuming and costly. We had to assemble a highly interdisciplinary team, forge partnerships with leading academic and clinical institutions, and relentlessly innovate to stay ahead.

One of our first major hurdles was intellectual property. In August 2000, we filed a seminal patent application for an "Artificial Intelligence System for Genetic Analysis." When the U.S. Patent Office was ready to grant the patent, a critical mishandling by our law firm, Heller Ehrman, led to costly delays. This experience taught us the hard but vital lesson of securing a strong IP strategy and working with trustworthy legal counsel.

Despite these setbacks, we pressed forward, driven by our mission to improve health outcomes through technology. Over the years, Iris Biotechnologies evolved from a bold idea into a company at the forefront of precision medicine. We developed innovative diagnostic tools for complex diseases like breast cancer and later established Iris Wellness Labs to expand our reach.

Throughout this journey, my sense of wonder at the power of science never waned. Biotech is a rare field where intellectual curiosity, technological innovation, and humanitarian purpose intersect. It continues to attract some of the brightest minds and boldest thinkers all united by a drive to push the boundaries of what's possible.

For aspiring entrepreneurs in biotech, my advice is to stay deeply attuned to the scientific frontier, build a strong network of advisors and collaborators, and never lose sight of the human impact of your work. The road is challenging but immensely rewarding for those driven by a passion for discovery and a commitment to improving health and quality of life.

Above all, embracing biotechnology entrepreneurship means signing up for a mission-driven journey of lifelong learning and impact. It means being part of a community committed to solving some of the most complex and consequential challenges facing humanity. There are few pursuits more meaningful or more demanding than using science to improve lives.

The field of biotechnology is constantly evolving, with new discoveries and technological advancements emerging at a rapid pace. To succeed in this dynamic environment, entrepreneurs must cultivate a deep understanding of the science underlying their innovations while also staying attuned to market trends, regulatory changes, and shifts in the healthcare landscape.

Biotech entrepreneurship presents unique challenges: long development timelines, high capital requirements, and rigorous scientific and regulatory standards. Unlike software or consumer products, where a minimum viable product can often be launched quickly, biotech innovations typically require years of research, preclinical work, and clinical trials before reaching the market. This means biotech founders must master long-term strategic planning, milestone-based execution, and clear communication with investors.

The regulatory environment in biotech is another critical factor entrepreneurs must navigate. Agencies like the FDA in the United States play a crucial role in ensuring the safety and efficacy of new medical technologies. Understanding these regulatory pathways and building them into your development plans from the outset is essential. This often involves engaging regulatory experts early and maintaining proactive communication with oversight bodies throughout the process.

You must also evaluate key regulatory decisions early on. For example: Will your product require FDA approval, or can it be marketed as a Laboratory Developed Test (LDT)?

And how might industry lobbying influence the rules that apply to you? If competitors are spending millions on lobbying, you need to understand why and what it means for your business.

Collaboration is also key in the biotech world. Many breakthrough innovations emerge from partnerships between startups, academic institutions, and established pharmaceutical or medical device companies. As an entrepreneur, your ability to build and manage strategic partnerships can be as important as your scientific or technical vision.

Ethics and social responsibility take on heightened importance in biotech, given the direct impact of these technologies on human health and well-being. Entrepreneurs in this field must grapple with complex ethical questions around genetic engineering, data privacy, equitable access to healthcare, and more. Establishing a strong ethical foundation early and intentionally builds trust with patients, clinicians, regulators, and the public.

The potential for impact in biotech is enormous. Advances in gene therapy, precision medicine, regenerative medicine, and AI-driven drug discovery are opening doors to radically improved health outcomes and quality of life for millions. This is what makes biotech so inspiring it blends scientific innovation with life-changing potential.

However, with that potential comes responsibility. Biotech entrepreneurs must balance innovation and commercial ambition with a commitment to patient safety and scientific rigor. Overpromising or cutting corners can damage not just your company, but also the reputation of the entire field.

# Chapter 2
## The Biotechnology Frontier

As you consider launching a biotech venture, it's essential to assess your motivations, capabilities, and readiness. Do you have the scientific grounding to lead or guide development? Are you prepared for the capital intensity and long timelines? Do you have access to a network of expert advisors, collaborators, and potential funders?

If your answer is yes, biotech entrepreneurship offers an incredibly rewarding path. You'll work at the forefront of science, build solutions that improve and save lives, and contribute to shaping the future of medicine.

Remember, success in biotech rarely happens overnight. It demands resilience, adaptability, and an unwavering commitment to learning. New discoveries will emerge. Market conditions will shift. Your ability to evolve with them is what will set you apart.

As we begin this exploration, I encourage you to think boldly about the problems you want to solve and the difference you want to make. Biotech offers endless opportunities for innovation, purpose, and impact. With scientific insight, entrepreneurial drive, and ethical clarity, you have the power to help redefine what's possible in human health.

The journey of biotech entrepreneurship is not easy. But for those with the passion and perseverance to pursue it, it may be the most fulfilling and impactful endeavor one can choose. So let's dive in and discover what it truly takes to succeed in this demanding, dynamic, and deeply meaningful field.

# Chapter 3
# Building an Exceptional Team

One of the most important lessons I have learned in my entrepreneurial career is that people are everything. No matter how brilliant your technology or how disruptive your vision, the success of your venture ultimately hinges on the quality and cohesion of your team.

Surround yourself with individuals who share your values, complement your strengths, and bring diverse perspectives to the table. In interviews, look for key qualities like relevant experience, creative problem-solving, ability to thrive in uncertainty, collaborative spirit, and commitment to the mission.

At Iris Biotechnologies, I had the privilege of assembling an incredible team of scientists, clinicians, engineers, and business strategists from renowned institutions like UCSF, Stanford, MD Anderson, and UC Berkeley. Their collective expertise and dedication were instrumental in overcoming complex challenges and achieving our milestones.

Beyond recruiting top talent, it is equally important to foster a culture of continuous learning, open communication, and shared purpose. Invest deeply in your team's development through mentorship, cross-training, and clear growth pathways. Lead by example, demonstrating the values and work ethic you wish to see in your organization.

The highest-performing teams are aligned around a common vision, leverage one another's strengths, and operate with trust and psychological safety at their core. They are resilient in the face of setbacks and passionate about realizing the company's potential for impact. Building this kind of exceptional team takes concerted effort but the long-term returns are immeasurable.

For first-time founders, my advice is to be very selective in your early hires, seeking out people who will elevate the entire enterprise. Spend time deliberately shaping your culture and living your values. Prioritize transparency and feedback from the start, and create a work environment where people feel heard and empowered. Celebrate wins and candidly acknowledge challenges. Above all, recognize that your team is your most valuable asset and treat them with respect and care.

Building an exceptional team in the biotech industry presents unique challenges and opportunities. The interdisciplinary nature of the field requires bringing together individuals with diverse backgrounds from molecular biology and genetics to data science and clinical medicine. Each team member must not only excel in their area of expertise but also collaborate across disciplines and contribute to shared strategic goals.

When recruiting for a biotech startup, look for individuals who combine deep scientific knowledge with an entrepreneurial mindset. They should be comfortable with ambiguity and able to adapt quickly as new data emerges or market conditions change. Passion for the mission is crucial in the face of inevitable setbacks and long development timelines, it's this shared commitment to making a difference that will keep the team motivated and cohesive.

Cross-functional collaboration is essential. Create integrated teams that bring together R&D, product development, regulatory, and business strategy. This diversity of perspective fosters more innovative solutions and bridges the gap between scientific discovery and commercial application. Encourage regular interaction and knowledge-sharing across groups to build a culture of collaborative innovation.

In the early stages of a biotech startup, it's common to rely heavily on a network of advisors and consultants to supplement the core team's expertise. Choose these advisors carefully, looking for individuals who not only have relevant technical or industry knowledge but also understand the unique challenges of startup environments. Treat them as strategic partners, engage them meaningfully, set clear expectations, and involve them in key decision-making.

As you grow, be intentional about preserving the culture and values that fueled your early success. It's easy for the collaborative, innovative spirit of a small team to get diluted as the organization expands. Establish programs, rituals, and feedback mechanisms that reinforce your core principles and enable open communication and collaboration across departments and geographies.

Leadership in a biotech startup requires a delicate balance. You need to provide clear direction and make tough decisions while also empowering your team of highly skilled professionals to take ownership and drive innovation. Adopt a leadership style that blends vision and decisiveness with humility, active listening, and inclusive decision-making. Foster an environment where input from all levels is valued.

Developing talent should be a top priority. In the fast-moving world of biotech, continuous learning is not just desirable it's essential. Encourage your team members to stay current with the latest scientific literature, attend conferences, and pursue ongoing education. Explore formal professional development pathways, academic partnerships, or in-house training to ensure your team stays ahead of the curve.

Creating a diverse and inclusive team is particularly important in biotech. Different backgrounds and perspectives can lead to more innovative problem-solving and help ensure that your products and services are designed with a broad range of end-users in mind. This diversity should extend beyond just scientific disciplines to include gender, ethnicity, socioeconomic background, lived experience, and cognitive diversity.

Remember that building a great team is an ongoing process, not a one-time task. Regularly assess your team's strengths and weaknesses, and be proactive about addressing gaps. Use performance reviews, feedback loops, and strategic hiring to continuously evolve your team's capabilities. This might mean bringing in new hires, providing additional training, or reorganizing to better leverage existing talent.

Foster a culture of intellectual honesty and rigorous debate. In science-driven fields like biotech, it's crucial that team members feel comfortable challenging assumptions, pointing out potential flaws in experiments or analyses, and proposing alternative hypotheses. Create structured forums for critical discussion such as lab meetings, innovation reviews, or peer-led audits to embed this openness into your workflow. This culture of constructive criticism and open dialogue is essential for maintaining scientific integrity and driving innovation.

At the same time, this critical thinking should be balanced with a sense of shared purpose and mutual support. Biotech development often involves setbacks and failures experiments don't always yield expected results, clinical trials can fail, and regulatory hurdles can arise. Teams that operate with empathy, persistence, and collective resolve will weather these storms more effectively and emerge stronger.

Formal mentorship programs can help foster both skill development and cultural cohesion. Pairing junior team members with experienced colleagues accelerates learning, improves retention, and builds cross-functional understanding. These relationships can also help bridge silos between different departments, reinforcing a shared sense of mission and collaboration.

As your company grows, establish structured and transparent career pathways for your team members. In a startup environment, roles often evolve rapidly, which can be both exciting and challenging for employees. Conduct regular check-ins, set clear performance expectations, and co-create personalized development plans to ensure that your top talent sees a future for themselves within your organization.

Remember that compensation is essential, but it's not everything. In the competitive biotech talent market, you may not always be able to offer the highest salaries, especially in the early stages. However, you can compete by providing purpose-driven work, a compelling mission, real ownership opportunities, and a culture that prioritizes learning, collaboration, and well-being.

Finally, don't underestimate the importance of celebrating successes together as a team. In the often long and challenging journey of biotech development, taking the time to acknowledge and celebrate milestones whether it's publishing a key paper, achieving a critical experimental result, or securing a new round of funding can boost morale and reinforce the sense of shared purpose that binds your team together. Small rituals of recognition, team retrospectives, and shared storytelling can help maintain momentum and build cohesion.

Building an exceptional team in biotech is both an art and a science. It requires a keen eye for talent, a commitment to ongoing development, and the ability to create an environment where diverse individuals can come together to do their best work. When executed with intention and care, this team-building process becomes the cornerstone of breakthrough innovation and lasting impact in human health.

# Chapter 4
# Defining Your Market Strategy

Another critical factor that can make or break a young company is its go-to-market strategy. Having a breakthrough idea is one thing, but deeply understanding your target customers, market landscape, and competitive positioning is essential to successfully commercialize that idea.

Before launching Iris Biotechnologies, we conducted extensive market research and competitive analysis. We assessed the size and growth trajectory of the personalized medicine space, identified key players and potential partners, and honed in on the specific applications where our technology could add the most value. This foundational work informed every aspect of our business model and product development roadmap. A winning market strategy requires clear answers to questions like:

- Who are your core customers, and what are their most pressing needs?
- How does your offering uniquely address those needs compared to alternatives?
- What are the key trends, regulatory factors, and competitive forces shaping your market?
- What is your pricing and distribution model to effectively reach and serve customers?
- How will you build brand awareness, trust, and loyalty in your target market?

The sharper and more differentiated your answers to these questions, the better equipped you will be to allocate resources efficiently and adapt to changing market conditions. In the fast-moving world of biotechnology,

17

precision in market strategy must be balanced with agility to respond to new clinical insights, policy developments, and scientific breakthroughs.

For Iris Biotechnologies, our choice to focus on oncology, particularly breast cancer analysis and treatment, was driven by a combination of factors: the strong unmet clinical need, the availability of robust genomic data, the alignment with our team's expertise, and the massive potential market. By intentionally narrowing our focus to breast cancer, we were able to establish a strong value proposition and build credibility with key opinion leaders and partners.

As the company grew, we continued to refine our market segmentation and positioning, expanding into new therapeutic areas and diagnostic modalities based on a clear view of our unique strengths. We also invested heavily in market development activities like clinical education, advocacy engagement, and health economic evidence generation to help shape the overall ecosystem for precision medicine adoption. These efforts allowed us not only to build demand for our solutions but also to contribute meaningfully to the broader adoption of personalized medicine.

Defining your market strategy is an ongoing process that requires customer engagement, competitive vigilance, and a willingness to pivot as needed. It also requires a balance between focus and flexibility, between confidence in your vision and humility to learn from the market. Getting this balance right is both an art and a science, but it is well worth the effort.

For aspiring founders, I recommend devoting significant time upfront to pressure testing your assumptions about the market and your place in it. Seek out mentors and advisors with deep industry expertise, but also get out of the building and talk directly to potential customers and partners.

Listen actively, challenge your assumptions, and be open to unexpected insights. Be prepared to iterate based on what you learn while staying true to your core purpose. A strong market strategy is the foundation upon which everything else is built.

In the biotech industry, defining your market strategy comes with unique considerations. The "customers" in healthcare are often multi-faceted you may need to consider patients, healthcare providers, payers, and regulators in your strategy. Each of these stakeholders has different needs, decision-making processes, and value drivers that you'll need to understand and address.

It's also crucial to recognize the long development timelines typical in biotech. Your initial market strategy needs to account for how the landscape might evolve over the years it takes to bring a product from concept to market. This includes anticipating shifts in clinical practice, regulatory frameworks, and reimbursement models. This requires a deep understanding of scientific trends, emerging technologies, and shifting healthcare paradigms.

When defining your target market, consider not just the size of the opportunity but also the feasibility of accessing it. In healthcare, factors like reimbursement policies, clinical guidelines, and standard-of-care practices can significantly impact market adoption.

Your strategy should include plans for generating the clinical evidence and health economic data needed to support the uptake of your product. Proactively engaging with payers and health systems early on can help shape a reimbursement path aligned with your business goals.

Competitive analysis in biotech should look beyond just direct competitors. Consider alternative approaches to addressing the same clinical need, as well as products or technologies that might be in development. Also, be aware of the potential for disruptive innovations that could fundamentally change the treatment paradigm in your target area. Scenario planning and technology forecasting can be powerful tools to help you prepare for these eventualities.

Partnerships often play a crucial role in biotech market strategies. Consider early on whether you plan to commercialize your product independently or in partnership with larger pharmaceutical or diagnostic companies. Each path has its pros and cons and will significantly impact your resource needs, timeline, and potential market reach.

Strategic alliances can provide access to distribution networks, regulatory expertise, and commercial infrastructure that would otherwise take years to build. On the other hand, going it alone may allow you to retain greater control and value capture so choose your path carefully based on your long-term goals.

Pricing strategy in healthcare is complex and heavily scrutinized. You'll need to balance the value your product delivers with the increasing pressure for cost-effectiveness in healthcare. Consider engaging health economists early to help build the value story for your product.

Health technology assessment (HTA) agencies and payer organizations will look closely at your product's cost-benefit profile, so having a compelling and data-driven narrative around clinical and economic value is key.

Regulatory strategy is inseparable from market strategy in biotech. Your choice of initial indication, the specific claims you aim to make about your product, and your plans for future label expansions will all impact your regulatory pathway. These decisions should be made with both scientific and commercial considerations in mind. Engage with regulatory consultants early and consider seeking feedback from agencies like the FDA or EMA during pre-submission meetings to de-risk your development path.

Remember that in biotech, your market strategy isn't just about selling a product it's about advancing a new approach to healthcare. Education and advocacy often play a big role. You may need to invest in raising awareness about the underlying science, changing clinical practices, or even shaping healthcare policy to create a receptive market for your innovation. Thought leadership, peer-reviewed publications, and engagement with medical societies can all help position your product within the broader clinical and scientific ecosystem.

Lastly, don't underestimate the power of patient voices in shaping healthcare markets. Engaging with patient advocacy groups and incorporating patient perspectives into your product development and commercialization strategies can be hugely valuable.

Patients are not just end-users they are increasingly influential stakeholders in driving demand, guiding trial design, and influencing payer and regulatory decisions. Building authentic relationships with these communities can differentiate your company and product in meaningful ways.

Developing a robust market strategy in biotech is a complex undertaking, but it's absolutely critical for success. It requires a unique blend of scientific insight, commercial acumen, and healthcare system knowledge. It's about building bridges between science and the clinic, between innovation and access, between vision and execution. But get it right, and you'll be well-positioned to translate your scientific breakthroughs into real-world impact, improving patient outcomes and transforming healthcare delivery.

# Chapter 5
# The Intellectual Property Imperative

In knowledge-intensive industries like biotechnology and life sciences, a robust intellectual property (IP) strategy is essential for protecting innovation, securing competitive advantage, and unlocking value through partnerships and transactions. However, developing and executing this strategy can be a complex and costly undertaking with little margin for error.

At Iris Biotechnologies, safeguarding our IP was a top priority from day one. We engaged experienced patent attorneys to help us craft strong, defensible patents around our core technology platforms and applications. We also implemented rigorous trade secret policies and confidentiality agreements with all employees and research partners to further protect proprietary knowledge.

One of our key early IP milestones was the filing of a foundational patent for our *Artificial Intelligence System for Genetic Analysis* in 2000. In 2008, the U.S. Patent Office informed our law firm that the application was ready to be granted a thrilling moment that would have validated years of innovation. But instead of celebrating, we faced an IP crisis.

Our law firm at the time, Heller Ehrman, mishandled critical correspondence from the USPTO, resulting in a deemed abandonment of the application. Reviving it took years of litigation, significant financial resources, and created massive uncertainty at a pivotal moment for the company. Heller's malpractice cost Iris an estimated $100 million in damages, and the experience underscored just how high the stakes are when it comes to IP in biotech.

Up until that point, we had worked closely with our patent attorney, and all of our previous applications had been granted worldwide. In this case, however, we were specifically advised to wait and take no further action. Unbeknownst to us, the entire Menlo Park patent team left Heller Ehrman on the same day, leaving no one to receive or respond to the USPTO's communication. The firm never informed us. Adding to the chaos, Heller filed for bankruptcy on December 28, 2008 just months after the collapse of Lehman Brothers and the failure of Heller's merger with Mayer Brown. Mayer Brown was the law firm for Lehman Brothers and about 50% of the Wall Street firms. The fallout of this timing amid the 2008 financial crisis exacerbated an already devastating mistake.

This experience taught us a hard but invaluable lesson: your IP counsel is as critical as your science. In biotech, a company's patent portfolio is often its most valuable asset the foundation of defensibility, investability, and strategic leverage. Any missteps in filing, prosecution, or enforcement can have catastrophic consequences.

Another core lesson was the importance of aligning IP strategy with business strategy. Rather than filing opportunistically, we became far more disciplined building a portfolio that supported our commercial goals, geographic priorities, and valuation strategy. We focused not just on patent quantity, but on relevance, enforceability, and the potential to block competitors or enable key partnerships.

We also became more strategic in how we communicated our IP story to investors and partners. By presenting a cohesive narrative around our IP's strength, scope, and alignment with market needs, we secured more favorable deal terms and increased investor confidence. Our IP portfolio evolved from a technical asset into a central pillar of our broader value-creation strategy.

For other biotech founders, I cannot overstate the importance of investing in a sophisticated, forward-looking IP strategy from the outset. That means not only hiring top-tier patent counsel, but also becoming educated yourself on the fundamentals of patent law, data exclusivity, freedom to operate, and competitive landscaping. It means integrating IP planning into your R&D and commercialization timelines, and staying proactive about reviewing and evolving your portfolio as your business evolves.

The complexity of biological systems, the rapid pace of scientific advancement, and the high stakes of medical innovation all make IP strategy in biotech particularly nuanced. One key consideration is the interplay between patent protection and data exclusivity. In many jurisdictions, biologic drugs are eligible for extended periods of market exclusivity through regulatory pathways often providing stronger or more predictable protection than patents alone. A sophisticated biotech IP strategy must thoughtfully leverage both.

Developing and defending a strong IP portfolio is expensive and time-consuming, but it is non-negotiable for building a durable and valuable company. In an industry where innovation is the product, a sound IP strategy is not just a legal formality it's a competitive weapon and a cornerstone of enterprise value.

Another critical dimension of IP strategy in biotechnology is the evolving legal landscape surrounding patent-eligible subject matter. In recent years, court decisions particularly in the U.S. have introduced significant uncertainty around the patentability of certain biotech innovations, including gene sequences, diagnostic methods, and some AI-driven technologies. Navigating this shifting terrain requires patent applications that are scientifically robust and legally resilient, crafted to withstand intense scrutiny under today's more restrictive standards.

Freedom to operate (FTO) analysis is equally vital. Biotech is a densely patented field, and the risk of inadvertently infringing on existing IP is high. A single missed claim can derail years of development. Conducting comprehensive FTO analyses early and revisiting them regularly as your product evolves is essential. Developing design-around strategies, securing licenses, or even adjusting development trajectories may be necessary to avoid costly litigation or commercial delays.

Patents in biotech often build upon prior inventions in intricate and layered ways. This makes understanding the broader patent landscape and identifying opportunities for improvement patents, new uses, or novel combinations of existing technologies a powerful strategic tool. Doing so requires not only legal and commercial savvy but also deep scientific insight and creative problem-solving.

Collaboration is common in biotech whether with academic institutions, other biotech firms, contract research organizations, or public-private partnerships. Each collaboration introduces complex IP considerations. Questions of ownership, licensing, and the handling of background IP and jointly developed innovations must be clarified upfront through well-structured agreements. A lack of clarity here can lead to disputes that delay progress and diminish the value of your IP.

In addition to patents, trade secrets play a key role in biotech IP strategy. For certain innovations such as proprietary manufacturing processes, AI algorithms, or unpublished data analytics tools trade secret protection may offer more durable or practical advantages than patenting. However, this requires strong internal controls, access limitations, and formal policies to ensure enforceability.

Given the global nature of biotechnology, an international IP strategy is non-negotiable. Protecting your innovations means filing in multiple jurisdictions, each with its own legal standards, enforcement challenges, and strategic considerations. Being aware of international treaties, such as the Patent Cooperation Treaty (PCT), and the IP enforcement environment in key markets like the EU, China, and India is crucial for long-term success.

As biotech continues to converge with adjacent fields artificial intelligence, nanotechnology, robotics, synthetic biology new layers of complexity and opportunity emerge. A future-ready IP strategy must include cross-disciplinary expertise to protect innovations that sit at these intersections and to capitalize on emerging markets and regulatory frameworks.

Importantly, intellectual property is not just a defensive shield it's a strategic asset for value creation. A strong IP portfolio can attract investment, secure lucrative partnerships, create barriers to entry for competitors, support pricing and reimbursement negotiations, and generate revenue through licensing or acquisitions. It is the engine that drives commercial traction in an industry where the product may take years to reach market.

In conclusion, while the Heller Ehrman incident was a painful chapter in Iris Biotechnologies' journey, it offered an unforgettable lesson: in biotech, a single legal oversight in IP can jeopardize everything. Vigilance, strategic foresight, and reliable legal counsel are not optional they are foundational. For biotech entrepreneurs, treating intellectual property as a core business priority not just a legal formality can mean the difference between success and failure.

By investing early in sophisticated IP strategy, multidisciplinary planning, and strong governance, companies can safeguard their innovations, multiply their enterprise value, and ultimately fulfill the promise of transforming lives through science.

# Chapter 6
# Securing the Right Capital

Funding is the fuel that powers the engine of entrepreneurship and nowhere is that more evident than in capital-intensive sectors like biotech, where innovation demands both time and significant resources. But not all funding is created equal. The choices you make about when, how, and from whom to raise money can have profound implications for your company's trajectory sometimes determining not just growth, but survival.

At Iris Biotechnologies, we were fortunate to have a strong network of supporters and a compelling vision that attracted interest from a diverse range of investors. In our early days, we relied primarily on angel investors and strategic partners individuals who were aligned with our mission and willing to bet on our team before we had fully proven ourselves.

These investors provided not just financial capital but also critical intangible resources: seasoned advice, industry connections, and early validation of our mission. They understood the long product development timelines in our industry and were patient in their expectations for returns. Their support gave us the runway to focus on building a strong scientific foundation and generating robust proof-of-concept data, a luxury not all startups in biotech can afford.

As we hit key technical and clinical milestones, our capital needs naturally grew. At that stage, we began preparing to take the company public, a major strategic shift that required a different kind of investor. We prioritized alignment over hype. We looked for partners who shared our values and long-term vision, rather than chasing the highest valuation or fastest term sheet. We were deliberate in how we staged each round of financing, seeking to minimize dilution and maintain strategic flexibility.

This measured approach paid off. By diversifying our funding sources and staying disciplined about when and how much to raise, we were able to maintain a healthy balance sheet and protect a strong equity position for our employees and early backers. Preserving equity isn't just about ownership. It's about motivation, retention, and culture.

For other founders considering the fundraising journey, my advice is to start with clarity. Get crystal clear on your milestones, timelines, and capital requirements. Be realistic about how much you need to raise to reach meaningful inflection points, and always build in a cushion for unexpected setbacks because they will come.

Next, carefully evaluate what kind of investor is the best fit for your current stage and long-term goals. Angel investors can be great partners in the early days when you need believers who can move fast and aren't scared off by ambiguity. Venture Capitalists can be powerful allies when it's time to scale, but be sure to do your diligence on their reputation, track record, and value-add beyond just money. Ask former portfolio companies not just how the VC behaved during the highs, but how they responded during the lows.

Strategic corporate investors can provide invaluable domain expertise and commercial insights but may come with complex strings attached. Make sure you understand those terms and the potential long-term consequences before taking their capital.

In the end, fundraising isn't just about cash. It's about building the right coalition of believers; people who will stand with you through the uncertainties of innovation and help you turn vision into reality.

When engaging with investors, focus on building genuine relationships, not just delivering transactional pitches. Too often, founders treat investor meetings like one-off sales presentations. But long-term capital partnerships thrive on authenticity, mutual respect, and shared conviction. The best investors are those who take the time to truly understand your business, provide honest feedback, and stick with you through the ups and downs. They don't just write checks, they challenge assumptions, open doors, and help you level up as a leader.

They can be force multipliers for your success, but only if there's a foundation of trust and alignment. That foundation starts early, often before a term sheet is even discussed. Communicate clearly, follow through on your commitments, and be transparent about both your vision and your challenges. Investors don't expect perfection, they expect integrity and adaptability.

Remember that fundraising is a means to an end, not the end in itself. Don't get so caught up in the chase for the next round that you lose sight of the fundamentals of building a great business. Product, team, traction, and culture - these are the true engines of value. The most successful fundraising efforts flow naturally from a company's strength and momentum, not from desperate quick fixes. If you build something truly valuable, capital will find you.

Finally, pace yourself. Fundraising is a marathon, not a sprint. Choose your partners wisely, and always prioritize long-term alignment over short-term gain. When done right, fundraising becomes more than just financing. It becomes a strategic advantage that propels your mission forward.

In the biotech sector, funding considerations take on added complexity due to the long development timelines, high capital requirements, and unique risk profile of life science ventures. Here are some additional points to consider:

1. Non-dilutive funding: In biotech, there are often opportunities for non-dilutive funding through government grants, research contracts, and partnerships. These can be excellent ways to advance your research and development without giving up equity. Programs like the NIH's Small Business Innovation Research (SBIR) grants can be particularly valuable for early-stage companies.

2. Staged financing: Given the long development timelines in biotech, it's common to raise money in stages tied to specific milestones. This allows you to minimize dilution in the early, high-risk stages and raise larger amounts at higher valuations as you de-risk the technology.

3. Crossover investors: As you approach later stages of development, consider engaging with crossover investors - those who invest in both private and public companies. These investors can provide a bridge to the public markets and help prepare your company for an eventual IPO.

4. Patient capital: Look for investors who understand the long timelines in biotech and are willing to be patient. Some specialized life science venture funds are structured with longer fund lifetimes to accommodate the extended development cycles in this industry.

5. Regulatory expertise: Investors with experience navigating the complex regulatory landscape in healthcare can be invaluable partners. They can help you anticipate and prepare for regulatory hurdles, potentially saving you time and money.

6. Syndication: In biotech, it's common to form syndicates of investors for larger rounds. This can help spread the risk and bring together complementary expertise. However, be mindful of potential conflicts or competing agendas among your investor group.

7. Alternative financing structures: Consider exploring alternative financing structures like royalty financing or venture debt. These can be useful tools for extending your runway without further dilution, especially as you approach commercialization.

8. International investors: Given the global nature of the biotech industry, don't limit yourself to domestic investors. International investors, particularly from biotech hubs in Europe and Asia, can provide not just capital but also valuable connections and insights into global markets.

9. Strategic alignments: When considering strategic corporate investors, think carefully about how their involvement might impact future partnering or M&A opportunities. While their industry expertise can be invaluable, make sure any deal terms preserve your flexibility.

10. Public markets: For some biotech companies, going public can be an attractive option for accessing larger pools of capital. However, this comes with increased scrutiny and reporting requirements. Carefully weigh the pros and cons and ensure you have the infrastructure to operate as a public company before pursuing this route.

11. Investor education: Many investors, even sophisticated ones, may not fully understand the intricacies of your technology or the specific challenges of your subsector. Be prepared to invest time in educating potential investors about your science, your market, and the broader context of your work.

12. Valuation considerations: Valuing early-stage biotech companies can be challenging due to the long time horizons and binary nature of many outcomes (e.g., clinical trial success or failure). Work with your advisors to develop a robust and defensible valuation model that accounts for the unique aspects of your technology and market opportunity.

13. Cash management: Given the high burn rates typical in biotech R&D, sophisticated cash management is crucial. Consider working with financial advisors who have specific experience in biotech to help you optimize your cash deployment and runway extension strategies.

14. Investor synergies: Look for investors who can offer synergies with your business beyond just capital. This might include access to specialized facilities, connections to key opinion leaders in your field, or expertise in areas like manufacturing or commercialization that complement your internal capabilities.

Funding continuity: In biotech, it's critical to always be thinking several steps ahead in your funding strategy. Start preparing for your next round even as you close your current one. This ensures you have the runway to achieve meaningful milestones and aren't forced into unfavorable terms due to cash constraints. Remember, at the end of the day, the fuel you need to power through the entrepreneurial journey is only partly financial. Equally vital is the fuel of passion, perseverance, and purpose - the internal drive that sustains you through setbacks, pivots, and long development cycles. By staying true to your mission, taking care of your team, and making consistent progress, you will attract the right capital at the right time to help you reach your destination.

In biotech, perhaps more than in any other field, the alignment between your scientific vision, business strategy, and funding approach is critical. The right investors can provide not just the capital to advance your R&D, but also the strategic guidance, industry connections, and long-term support necessary to navigate the complex path from scientific discovery to market impact. This alignment becomes your compass guiding not only what you build, but also who you build it with.

As you build your funding strategy, always keep the ultimate goal in mind: to translate scientific breakthroughs into therapies or technologies that can improve human health. This mission-driven approach, combined with a sophisticated understanding of the unique funding landscape in biotech, will position you well to secure the capital you need to bring your innovations to life. Raising money should never feel like a detour from your mission. It should be a reflection of it.

Raising capital usually starts with a business plan. Crafting a strong business plan is both an art and a discipline. It forces you to crystallize your thinking around product-market fit, regulatory strategy, competitive landscape, and financial runway. It's difficult to write a great one, but when you do, you'll know it. A solid business plan doesn't just impress investors; it clarifies your own vision and helps you make better decisions.

Having served as a judge in the Global VCIC, the world's largest venture capital investment competition with over 120 university and graduate school teams competing, I've seen firsthand how powerful it is for founders to practice telling their stories. VCIC is the only platform where students step into the shoes of Venture Capitalists (VCs) for the day, while real startups gain exposure and feedback to accelerate their fundraising process.

My advice: seek out opportunities to rehearse the fundraising experience before the stakes are high. One such venue is Santa Clara University's California Entrepreneurship Program, where I've served as a mentor. These practice rounds help refine your pitch, stress-test your assumptions, and expose you to the questions and concerns real investors will raise - all in a lower-pressure environment.

In the high-stakes world of biotech entrepreneurship, knowledge is power, but preparation is momentum. Take the time to hone your narrative, build authentic investor relationships, and align every dollar you raise with your long-term vision. That's how real progress happens and how real impact begins.

# Chapter 7
# Starting Small and Government's Role

My professional entrepreneurship journey has centered on medium-sized ventures with the potential to scale into major enterprises. In many ways, it mirrored the humble yet bold beginnings of tech giants like Apple. On April 1, 1976, Steve Jobs, Steve Wozniak, and Ronald Wayne founded Apple Computer Company, registering it as a California business partnership.

The first Apple computer was designed in Wozniak's apartment and built in Jobs' garage. Their original headquarters was in Cupertino, California, just adjacent to Saratoga, where I started Iris Biotechnologies in the most personal of spaces: my living room.

In the early days, one of my employees worked out of my garage, which wasn't a traditional workspace. It was a unique hybrid of an office and a personal library, lined wall to wall with bookshelves holding about a thousand books, DVDs, and CDs. Another 1,500 books were stored throughout the house. I never used the garage for parking; instead, I set it up with tables, a combination lock on the door, and the freedom for her to come and go independently except during scheduled meetings. That makeshift setup was our first lab, our first office, and our first step toward building something meaningful.

First and foremost, anyone considering entrepreneurship needs to be crystal clear on his or her purpose. Why do you want to start a company? What is the impact you hope to make? What relevant knowledge or experience do you bring to the table? And perhaps most importantly, can you attract the right people - those who not only believe in your

37

mission but also can help you secure the resources your business needs to thrive?

If you're entering a highly competitive space like biotech, for example, it's vital that both your family and your partners' families understand the risks involved. They should be prepared for the significant demands in time, capital, and emotional bandwidth. Entrepreneurship isn't just a professional decision; it's a personal and relational one, too.

When it comes to forming a company, one of your first decisions is choosing the legal structure. Most people assume a C corporation is the default, but that's not always the best option. From a tax standpoint, an S corporation may offer advantages, especially for smaller, closely held companies. Similarly, deciding whether to incorporate in California or Delaware can have long-term legal and financial consequences.

You also need to consider whether a corporation or a limited liability company (LLC) best suits your goals. Before hiring a lawyer to handle your incorporation, take the time to educate yourself. The more you understand upfront, the better positioned you'll be to make informed, strategic decisions.

Legal costs in business can be staggering. Corporate matters, patent filings, enforcement actions, and intellectual property disputes can run into the millions. Even defending against frivolous lawsuits can drain your resources because whether you're right or wrong, you'll still have to pay legal fees to protect your company.

And remember, corporations cannot represent themselves in court. Even if you're knowledgeable enough to draft your own filings or mount a strong legal argument, the law requires that a licensed attorney represent the corporation. That's another reason to budget wisely and build strong legal relationships early on.

It can take years, sometimes a decade or more, before a patent application is granted, if it's granted at all. So, it's critical that your patent attorney is not just competent but also trustworthy, detail-oriented, and financially stable enough to stay with you for the long haul. Patent law varies widely across jurisdictions, the U.S., European Union, Japan, and other nations each have distinct standards and review processes. What gets approved in one country may be narrowed or rejected in another.

Make sure your patent claims are enforceable, not just granted. It's not enough to have a patent on paper; it has to stand up to legal scrutiny if challenged. And here's the nuance many entrepreneurs miss: the attorneys who specialize in helping you obtain patents, known as patent prosecution attorneys, are often not the same ones who specialize in enforcing them in court. These are two very different legal disciplines. Choose your words carefully during the application process, and consult a seasoned patent litigation or enforcement attorney when drafting claims that may one day be tested. Sometimes, a single word can be worth millions, or cost you just as much.

We spent over $1,000 per hour on our corporate attorney and slightly less on our patent attorney. That kind of billing forces efficiency. We tried to do as much groundwork as possible before involving the lawyers, drafting documents, gathering data, outlining claims. The goal was to enter every legal meeting informed, prepared, and specific in our needs. Good legal advice can be invaluable, but poor or uninformed counsel can derail your startup or cost you the company. Choose your advisors wisely and get second opinions when needed.

Some entrepreneurial ventures push the boundaries of innovation so far that they require government involvement, strategic diplomacy, or international cooperation. These aren't your typical startups, they're global-scale initiatives with geopolitical implications.

Take, for example, the New Silk Road, officially known as China's Belt and Road Initiative (BRI). This massive infrastructure and development program, launched in 2013 by President Xi Jinping, was originally conceived to connect East Asia and Europe via a network of roads, railways, and maritime routes. But by 2024, the BRI had expanded its scope and influence to Africa, Oceania, and Latin America, reshaping global trade corridors and extending China's economic reach. As of 2025, more than 150 countries have signed on to participate in some form.

While China promotes the BRI as a win-win development effort, many analysts in the West and in Asia view it with caution. Some consider it a geopolitical Trojan horse, a mechanism for extending Chinese influence under the guise of economic partnership. Ballooning costs, mounting debt in recipient nations, and shifting public sentiment have led to growing opposition in certain regions. The United States shares concerns that the BRI could enable China's military and political expansion in strategic regions.

Still, there is no doubt that this trillion-dollar initiative will change the world. The hope is that it will uplift billions by improving infrastructure, connectivity, and economic opportunity, though the long-term effects remain to be seen.

This tension between invention and implementation is nothing new. Historically, the Chinese were the original inventors of transformative technologies such as paper and woodblock printing. But it was Johannes Gutenberg's invention of the movable-type press in Europe that revolutionized information distribution and sparked the global spread of books, including the Bible.

Similarly, the Chinese pioneered gunpowder and early cannon technology. Yet it was the Turks who turned it into a formidable military force, using it to build the Ottoman Empire. At its peak, this empire stretched across Southeast Europe, West Asia, and North Africa, dominating trade, culture, and politics from the 14th to the early 20th centuries. From the early 1500s to the 1700s, it even controlled parts of Central Europe. The empire eventually crumbled following World War I, but its legacy lives on in modern geopolitics.

Fast forward to today, the United States was the birthplace of the Internet, a defining technology of our era. American companies like Apple, Microsoft, Google (Alphabet), Facebook (Meta), Amazon, and Nvidia continue to shape the digital landscape. On the other side of the Pacific, China has rapidly developed its own tech giants, including Alibaba, Tencent, ByteDance, Baidu, and Huawei.

TikTok, owned by Beijing-based ByteDance, became the most downloaded app of all time in 2024, underscoring China's growing dominance in digital media and global influence.

Legal frameworks have played a key role in enabling this digital transformation. On October 21, 1998, President Bill Clinton signed a groundbreaking bill that imposed a ten-year moratorium on discriminatory and multiple taxation of the Internet and electronic commerce. This policy fostered innovation and encouraged early growth in the tech sector. Congress extended the moratorium several times before making it permanent on February 24, 2016. Soon after, the European Union implemented similar protections, helping to create the global e-commerce ecosystem we know today.

As a result of the U.S. government's Internet tax relief policies, tech companies like Amazon were able to grow rapidly, overtaking traditional bookstores and retailers. Amazon began by selling books online, but its business model evolved. Today, Amazon operates out of vast, strategically placed warehouses and sells almost every consumer product imaginable. In addition to its massive retail presence, the company also dominates the cloud services industry through Amazon Web Services (AWS), offering secure, reliable, and scalable solutions for websites, streaming, big data, and artificial intelligence platforms.

On January 26, 2024, Amazon had a market value of $1.64 trillion. In comparison, Apple and Microsoft were valued at $2.98 trillion and $3.00 trillion, respectively. Alphabet (Google's parent company) stood at $1.91 trillion, and Meta (formerly Facebook) at $1.01 trillion. These valuations reflect a dramatic consolidation of digital power and wealth in just a few decades, driven by technological innovation and global demand.

Artificial Intelligence (AI) is now the biggest technology race among these companies. Nvidia, a pioneer in graphics processing units (GPUs), has emerged as the global leader in AI chip manufacturing. Its GPUs are essential for powering the machine learning systems behind everything from ChatGPT to autonomous vehicles. Nvidia became the third publicly traded U.S. Company to reach a $2 trillion market valuation, following Apple and Microsoft.

Meanwhile, Tesla Motors revolutionized the electric vehicle (EV) market, transforming EVs from novelty items into aspirational products. Founded in 2003, Tesla introduced its first vehicle, the all-electric Roadster, in 2008. Elon Musk, who joined as chairman and later became CEO, has been the company's public face and chief innovator. As of January 26, 2024, Tesla's market capitalization stood at $574.21 billion.

In 2023, Tesla delivered 1.81 million vehicles, representing a 38% year-over-year increase. Its production output grew by 35% to 1.85 million units. This rapid scale-up was supported by Tesla's gigafactories across the U.S., Germany, and China, which help meet global demand. However, Tesla is now being rivaled by China's BYD (Build Your Dreams), which shipped its first electric car in 2009. BYD, backed in part by Warren Buffett's Berkshire Hathaway, sold approximately 1.57 million EVs in 2023 a 73% increase from the previous year.

In 2024, BYD narrowly overtook Tesla in production numbers, with 1,777,965 electric vehicles manufactured, compared to Tesla's 1,774,442. This milestone marks a significant power shift in the global EV market, highlighting China's growing industrial strength in clean energy technologies.

Government subsidies have played a pivotal role in boosting consumer adoption of EVs, especially in the United States. The average direct federal subsidy per EV over ten years is nearly $9,000. When including local utility company incentives and rebates, the total subsidy can exceed $10,000 per vehicle. These policies aim to reduce greenhouse gas emissions and accelerate the transition to a more sustainable transportation system.

The journey to electric mobility was built on the back of combustion engine innovation. In 1872, American engineer George Brayton invented the first liquid-fuel internal combustion engine. Four years later, in 1876, Nicolaus Otto, Gottlieb Daimler, and Wilhelm Maybach developed and patented the compressed-charge, four-stroke cycle engine, which became the blueprint for modern automotive engines. In 1879, Karl Benz patented a two-stroke gas engine, laying the groundwork for the automobile era. These early breakthroughs ushered in the Age of Automobiles and the global reliance on petroleum-based fuels.

Oil, often referred to as "Black Gold," is one of the world's most valuable and strategic natural resources. It powers cars, planes, ships, and industrial machinery. But oil's impact goes beyond energy. It is a raw material in countless products, from plastics and pharmaceuticals to paints, fertilizers, detergents, and even cosmetics.

Before the mass extraction and refining of crude oil, whale oil was one of the primary sources of fuel, especially for lamps. Whale oil burned cleanly and without much odor, making it highly desirable. It was also used in lubricants, soaps, textiles, varnishes, explosives, and paints. However, as whale populations dwindled and costs rose, the industry declined, leading to the rise of petroleum as a cheaper, more scalable alternative.

Nantucket, Massachusetts, was once the epicenter of the American whaling industry. One of its most famous whaling ships, the Essex, was rammed and sunk by a sperm whale in 1820. That true story inspired Herman Melville's iconic novel Moby-Dick. I enjoyed reading that book and visiting the Nantucket Whaling Museum, a powerful reminder of how human ambition and nature collide in history. I also enjoyed a trip to Martha's Vineyard just 38 miles (63 km) from Nantucket and viewing the Kennedy Compound from the sea.

The Kennedy Compound, located on Cape Cod along Nantucket Sound, consists of three houses spread over six acres (2.4 hectares) of waterfront property. It was once the summer home of Joseph P. Kennedy, an American businessman, investor, and U.S. Ambassador to the United Kingdom. He lived there with his wife, Rose, and their children, including U.S. President John F. Kennedy and U.S. Senators Robert F. Kennedy and Edward M. Kennedy. The compound remains an enduring symbol of American political legacy and public service.

Interestingly, in the 1860s, U.S. government policy favored kerosene as a cleaner and cheaper lighting alternative, keeping taxes on it low. This strategic move accelerated the decline of whale oil, marking a pivotal moment in America's transition to fossil fuels. This was one of the earliest examples of how government incentives and tax policies could dramatically reshape entire industries.

In refining crude oil to make kerosene for lighting, gasoline was discovered as a by-product. Initially, it was discarded as waste. However, due to its ability to vaporize at low temperatures, gasoline later became a revolutionary fuel for engines. At a depth of almost 70 feet (21 m), Edwin L. Drake struck the first oil well in the United States in 1859 near Titusville, Pennsylvania, marking the birth of the modern petroleum industry.

Gasoline became a necessity in the automotive industry after Nikolaus Otto developed the four-stroke internal combustion engine in 1876. This invention not only shaped transportation but also set the stage for mass industrialization across sectors. Today, almost all gasoline is used to fuel cars, while a small portion powers agricultural equipment, lawnmowers, boats, and smaller piston-engine planes.

Fossil fuels like petroleum supply more energy to the world today than anything else. About 45% of petroleum becomes gasoline. Other lighter chemicals derived from refining include natural gas, liquefied petroleum gas (LPG), jet fuel, and kerosene. These lighter fractions fuel everyday energy needs such as cooking, flying, and home heating.

A lot of the heavier products go into lubricants, plastics, and asphalt. One of the byproducts of oil refining is petroleum coke, which is important for fertilizer production. Over half of the world's known crude oil reserves are located in the Persian Gulf basin. The US is the biggest oil consumer. Despite having significant domestic production, the U.S. still relies heavily on imports and global oil pricing.

The consequence of the U.S. government promoting kerosene through low taxation ushered in the era of oil dominance. Increased subsidies for oil especially for gasoline dramatically impacted the earth in both good and bad ways. In addition to providing affordable energy for heating, transportation, and electricity generation, oil also pollutes the environment and contributes to climate change. Greenhouse gas emissions from burning fossil fuels are a leading cause of global warming, rising sea levels, and severe weather events.

It's convenient to use plastics, but plastic waste is a major environmental and health problem. Fertilizers and pesticides, also made from oil, are widely used in agriculture. These chemicals increase crop yield but can also contaminate soil and water, disrupt ecosystems, and affect human health.

"Who Pays the Price: The Real Cost of Fossil Fuels" was the opening statement delivered by U.S. Senator Sheldon Whitehouse (D-RI), Chairman of the Senate Budget Committee, on May 3, 2023. In 2022, fossil fuel subsidies hit a record $1 trillion globally, while Big Oil companies earned $4 trillion. This stark contrast highlights the imbalance between public spending and corporate profit, raising serious ethical and environmental questions.

Many farmers choose to use herbicides, insecticides, and fungicides to protect their crops and enhance soil nutrients. Pesticides are chemical and organic mixtures, including insecticides, fungicides, and plant growth regulators. Herbicides are designed for plants. Pesticides kill insects, rodents, and fungi. Fertilizers are plant nutrients added to the soil to promote growth. However, overreliance on these chemicals has led to pesticide-resistant pests and nutrient depletion in soil, prompting the need for more sustainable agricultural practices.

Historically, the Queen owned the most land in England. The Church owned the second most. Neither the Queen nor the Church worked for the land, but commoners accepted it. This long-standing tradition of inherited wealth and unequal land distribution has shaped class divisions and property laws that still persist today.

Many nations have gone to war because of oil. Before oil, nations invaded others for land, spices, and commodities like gold. Oil has become the modern strategic resource, driving foreign policy, trade agreements, and military action in regions like the Middle East.

It is interesting to note that the world's richest oil and gas reserves are centered in the region of the Garden of Eden mentioned in the Bible. The Middle East, especially the Persian Gulf, is a hub for oil and gas. This convergence of spiritual lore and natural resources has added geopolitical and cultural complexity to the region's modern identity.

Tree trunks, leaves, twigs, and roots become coal, and oil is formed from ancient algae and cyanobacteria under immense heat and pressure. Oil is primarily made of carbon and hydrogen. Methane, $CH_4$, is the simplest hydrocarbon. Oil components like octane ($C_8H_{18}$) contain larger hydrocarbon molecules. With too much underground heat, oil breaks down to make methane. These processes span millions of years, underscoring the non-renewable nature of fossil fuels.

There was a time and a place where people valued salt more than gold, and people simply lived on land. Now, people trade away most of their lives working for a tiny piece of land. One in three U.S. residents never owns a home. Housing has become one of the most visible symbols of inequality, with rising prices and stagnant wages pushing ownership out of reach for many.

Why is there so much inequity and injustice in the world? Extremely wealthy people bribe corrupt government officials to get what they want whether it is money, power, or personal fulfillment. Corporate lobbying, tax havens, and monopolies further deepen this divide, allowing the rich to accumulate more wealth while the poor struggle to survive.

Should we ignore injustices that deprive both current and future generations? We live in an age of extreme inequality between the super-rich and the rest of society. Will this lead to severe instability around the world? History shows that extreme inequality often precedes social unrest, revolution, or economic collapse. If left unaddressed, the world may face mounting crises that transcend borders.

In the U.S., private sector regulation tends to be lighter. Therefore, fake news and extreme information are rampant. U.S. political divisions are getting worse. Technological platforms in Europe are becoming more regulated, and they have to take more responsibility for what they post. These differing regulatory approaches create an uneven digital landscape where misinformation can spread unchecked in some regions while being curtailed in others.

It is a balancing act between inventions and regulations. Inventions are important because they change history. AI is about to change the world in profound ways. But just like oil and plastic before it, AI will need ethical oversight to prevent harm. First, let's look at some of the major inventions in early human history.

Mesopotamia was the site of the earliest human civilization around 10,000 BCE. It inspired some of the most significant inventions from four to six thousand years ago, including many items that are taken for granted today.

The inventions include writing, the wheel, mathematics, time, mass-produced ceramics, cylinder seals and envelopes, mass-produced bricks, cities, maps, the sail, agriculture (the Plow) and irrigation, the concept of cartography, astrology, astronomy, chariots, metal fabrication, board games, soap, and the law system.

These represent only a small fraction of their technological, cultural, and scientific advancements. The Scholar Samuel Noah Kramer lists 39 'firsts' from ancient Sumer in Mesopotamia:

- The First School
- The First Case of 'Apple Polishing'
- The First Case of Juvenile Delinquency
- The First 'War of the Nerves'
- The First Bipartisan Congress
- The First Historian
- The First Tax Reduction Case
- The First 'Moses'
- The First Legal Precedence
- The First Pharmacopoeia
- The First 'Farmer's Almanac'
- The First 'Experiment in Shade-Tree Gardening'
- Man's First 'Cosmogony and Cosmology'
- The First Moral Ideals
- The First 'Job'
- The First Proverbs and Sayings
- The First Animal Fables
- The First Literary Debates
- The First Biblical Parallels
- The First 'Noah'
- The First 'Tale of the Resurrection'
- The First 'Saint George'
- The First Case of 'Literary Borrowing'
- Man's First Heroic Age
- The First Love Song
- The First Library Catalogue
- Man's First Golden Age
- The First 'Sick' Society
- The First 'Liturgy Laments'
- The First 'Messiah'
- The First 'Long-Distance Champion'

- The First Literary Imagery
- The First Sex Symbolism
- The First 'Mater Dolorosa'
- The First Lullaby
- The First Literary Portrait
- The First 'Elegy'
- Labor's First Victory
- The First 'Aquarium'

Writing was the most influential invention in Mesopotamia, and they kept their records on millions of baked clay tablets known as cuneiforms. These tablets preserved the region's beliefs, history, and culture, which significantly influenced later civilizations. Cuneiform is the earliest known system of writing, enabling laws, trade, astronomy, and literature to be recorded and passed down. The Epic of Gilgamesh, one of the oldest surviving works of literature, originated from this tradition.

Mesopotamia is in the northern part of the Fertile Crescent in southwest Asia within the Euphrates and Tigris rivers. It is also known as the Cradle of Civilization. Mesopotamia includes present-day Iraq and parts of Iran, Kuwait, Syria, and Turkiye in the Middle East. Its fertile lands, formed by river flooding, allowed for abundant agriculture, which supported growing populations and the rise of cities.

The oldest writing form, Cuneiform Script, was created in Mesopotamia about 3200 BCE. All civilizations in ancient Mesopotamia used a cuneiform writing system until the alphabetical system was introduced in about 100 BCE. Cuneiform evolved from pictograms to complex symbols over time, enabling communication across trade, governance, and religion.

Approximately half a million to two million cuneiform tablets have been excavated in modern times, and only about 30,000–100,000 have been read or published. Mesopotamia's counting method was based on 60. That is where we got 60 minutes in an hour, 60 seconds in a minute, 12 months in a year, and a 360-degree circle. This sexagesimal system continues to influence our concepts of time and geometry today. The Sumerians also invented an elemental abacus between 2700 and 2300 BCE, demonstrating an early form of computational thinking.

The first map was developed in Mesopotamia, but Roman and Greek cartography was more advanced. The Greek philosophers developed the idea of a spherical Earth around 350 BCE, which later resulted in our modern world map. However, the Babylonians laid the foundation by creating detailed maps of land boundaries and star charts, revealing an advanced understanding of geography and astronomy for their time.

The Sumerians documented Mercury, Venus, Mars, Jupiter, and Saturn's movements. They accurately predicted the planets' movements long before the Greeks, the Mayans, and other civilizations did thousands of years later. They developed a lunisolar calendar and could forecast eclipses, linking celestial events to religious and agricultural cycles.

The Greeks later absorbed astronomical concepts of patterns like Sagittarius, Leo, and Capricorn from the Sumerians and Babylonians. Ancient Mesopotamia used constellations to mark crop sowing and harvesting times. They also mapped movements in the sky, the moon, the stars, and the Sun to foretell cosmic events like an eclipse. This celestial knowledge was foundational for the development of zodiac astrology, influencing later Egyptian, Persian, and Greek systems.

The people of Mesopotamia invented 2, 4, and 6-wheel chariots that transformed ancient transportation and warfare used by different civilizations and nations. These innovations were critical for military strategy, allowing faster movement and tactical advantage in battle.

They invented large cities for the first time in human history. This was made possible by their agricultural technologies, improved transportation, mass-produced ceramics and bricks, the potter's wheel for making clay pots, and other inventions. Uruk, one of the earliest cities, had complex infrastructure, including temples, administrative buildings, and defensive walls.

Their invention of cities was both a gift and a burden in our modern world. Along with the conveniences, the cities had more crimes and other conflicts that were dealt with by their invention of a law system. The Code of Hammurabi, written around 1754 BCE, is one of the earliest and most complete written legal codes, promoting justice through publicly displayed laws.

Mathematics and time invented in Mesopotamia have been useful not only in finance, trades, and taxation but also in construction, engineering, medicine, and computer science. Around 2,000 BCE, Babylonian mathematicians created arithmetic, multiplication tables, square roots, division, and algebraic equations. They even solved quadratic and cubic equations, showing abstract reasoning and symbolic representation that predated Greek mathematics by centuries.

The Mesopotamians also played a vital role in sea travel through their sailboat invention around 5,500 BCE. Their early sailing vessels enabled long-distance trade along rivers and across the Persian Gulf, exchanging goods like grain, textiles, copper, and spices.

The first beer production was credited to the Sumerians in 4,000 BCE. Beer is the third most popular drink in the world, after potable water and tea.

Sumerians even had a goddess of beer, Ninkasi, and her hymn doubles as one of the oldest surviving beer recipes. It is one of the oldest alcoholic drinks in the world and has saved many lives in places without clean water.

In ancient times, fermented beverages were often safer to drink than untreated water due to alcohol's sterilizing properties.

The ancient Sumerians of Mesopotamia were the ancient equivalent of the people in Silicon Valley due to their prowess and passion for technological invention. They transformed how humans cultivated food, lived in cities, traveled, communicated, and kept track of information and time. They are the giants on whose shoulders we stand. Their innovations laid the groundwork for civilization as we know it, from written language to urban planning. We are grateful to the Sumerians for leading the way.

In our times, the San Francisco Bay Area, especially Silicon Valley, is the hub of entrepreneurship. Call and talk to some of the entrepreneurs and learn from them wherever you are.

Stanford University has produced many successful entrepreneurs, such as Elon Musk, founder of Tesla and SpaceX; Larry Page and Sergey Brin, founders of Google and Alphabet; and Jen-Hsun Huang, founder of Nvidia. According to Stanford, "A successful entrepreneur will possess many abilities and characteristics, including the ability to be curious, flexible and adaptable, persistent, passionate, willing to learn, a visionary and motivated."

Here is a company that some of you may not know about but that impacts many lives every day. Cisco Systems was founded in 1984. The company's headquarters are in San Jose, California. The founders of Cisco Systems were Leonard Bosack and Sandra Lerner, a married couple who met while students at Stanford University. In 1986, Cisco sold its first big success product, a router that served multiple network protocols.

In need of cash for expansion, the founders accepted funding from Sequoia Capital, a venture capital firm. Sequoia took control of the company in late 1987 and hired John Morgridge as president and CEO in 1988. Soon after Cisco's IPO (Initial Public Offering) in 1990, Lerner was fired. Bosack then quit, and their marriage ended the same year. This story highlights the critical importance of choosing investors and partners wisely. Entrepreneurship is as much about relationships and trust as it is about innovation.

Cisco's market valuation peaked on March 24, 2000, at more than $500 billion, surpassing Microsoft as the world's most valuable company. It was valued at less than $200 billion on March 20, 2024. The rise and fall of Cisco's valuation illustrate the volatile nature of tech markets and the need for entrepreneurs to stay adaptable and resilient.

As an entrepreneur, you must choose your investors wisely. It can be very costly, personally and financially.

Heller Ehrman was the premier law firm in the life science industry when I hired them to do my company's corporate and patent work in biotechnology. At its peak, Heller Ehrman, an international law firm, had more than 730 attorneys in 15 offices across the United States, Europe, and Asia. We were happy with their corporate work for eight years. We were content with their patent work until we discovered their $100 million malpractice against our company.

They put us in their corporate marketing literature as a company they proudly represented in our industry. Unfortunately, after their merger with another law firm, their ethics deteriorated. Our corporate lawyer and patent lawyer both left the firm.

Our patent attorney, who earned his PhD in Chemistry from UCLA and JD from Santa Clara University, successfully prosecuted all of our patent applications worldwide to become patents. I trusted him. After he and all the patent attorneys in their Menlo Park, CA office left the firm on the same day, Heller committed multiple malpractices and concealed them. This caused $100 million in damages to us. This experience was a harsh lesson not taught in MBA school but critical for anyone navigating the complex interplay of law, business, and innovation.

Because I lost confidence in our corporate attorney and was deeply troubled by Heller Ehrman's post-merger issues of greed and ethical lapses, I made the difficult decision to hire another law firm based in New York to take Iris Biotechnologies public. We successfully began trading publicly in August 2008 - a challenging time, right near the peak of the Great Recession.

To build a world-class team, I recruited PhDs in biology, chemistry, and chemical engineering, along with MDs, MBAs, and other experts from prestigious institutions including UC Berkeley, Parke-Davis (now Pfizer), UCSF, Stanford, UCLA, Caltech, Case Western, MD Anderson, and many more.

Our shareholder base reflected this commitment to excellence. It included graduates from UC Berkeley, Harvard, Yale, UCSF, Stanford, Caltech, Santa Clara University, UC Davis, University of North Carolina, and other leading schools across diverse fields. Business owners from various industries around the country also invested in Iris.

Although we were offered venture capital funding, I declined. I believed in maintaining control and integrity over our company's direction, especially after the painful experience with Heller Ehrman.

Unfortunately, we were blindsided by Heller Ehrman's $100 million malpractice and their repeated concealment of the issue. Disturbingly, Heller knew they owed us money when they filed for bankruptcy protection. Iris was listed among the unsecured creditors in Heller's bankruptcy filing, but they never informed us. Here are the details of what transpired in bankruptcy court.

*Iris vs. Heller*

**February 11, 1999** Mr. Simon Chin hired Mr. Bruce Jenett at Heller Ehrman, a prestigious, century-old law firm with over 600 attorneys, to incorporate Iris Biotechnologies Inc. and manage all legal matters for the company.

**August 17, 2007** Following Heller Ehrman's merger with another firm, which brought significant internal turmoil marked by greed, competence issues, and ethical concerns, Iris lost confidence in both Heller Ehrman and its corporate attorney, Mr. Bruce Jenett, to advise on taking the company public. Iris subsequently engaged a New York law firm to handle the public offering. In response, Mr. Jenett issued an illegal disengagement letter to Iris. Under U.S. patent law, a law firm cannot abandon a client without valid cause. The U.S. Patent and Trademark Office (USPTO) denied Heller Ehrman's petition to disengage, meaning the firm legally remained Iris's patent attorney from February 1999 through September 2011.

**December 14, 2007** Heller Ehrman successfully prosecuted all of Iris's patent applications worldwide to date. Dr. James A. Fox, the patent attorney handling Iris's portfolio, assured Mr. Chin, "Heller will send out a contemporaneous email whenever Heller forwards files in the future." Trusting Dr. Fox's deep familiarity with Iris's cases and their strong working relationship, Mr. Chin followed his advice to simply wait.

**December 20, 2007** Mr. Chin wrote to Dr. Fox:

*"It appears that Heller Ehrman has evolved into a firm that has problems not only with greed but also with competence and ethics. What happened has nothing to do with you, and I remain respectful of you and consider you a friend."*

***February 2008 – Mr. Bruce Jenett left Heller Ehrman.***

**Mar. 4, 2008** – Dr. Fox and the rest of the patent attorneys in Menlo Park left Heller Ehrman on the same day. Dr. Fox and Heller Ehrman did not inform Iris, and Iris was unaware of this exodus.

**March 21, 2008** – Heller Ehrman committed its first malpractice against Iris Biotechnologies by failing to forward a critical USPTO document that the firm had received. This failure led to the abandonment of Iris's most important foundational patent in the United States the world's most significant market.

*The U.S. Bankruptcy Court, under Judge Dennis Montali, later confirmed that Heller did indeed fail to forward crucial USPTO documents to Iris in March and October 2008. This malpractice caused a six-year delay in obtaining U.S. Patent No. 8,693,751, titled "Artificial Intelligence System for Genetic Analysis." Meanwhile, Iris's competitor surged ahead and was sold for $2.8 billion in 2019.*

**October 7, 2008** – Heller Ehrman committed a second malpractice by failing to forward the USPTO's notice of Iris's patent abandonment, which the firm had also received. This second malpractice concealed the first, effectively preventing Iris from discovering the wrongdoing and filing a timely malpractice claim while Heller still maintained malpractice insurance. *The bankruptcy court explicitly found that Heller did not forward this critical notice.*

**December 28, 2008** – Heller Ehrman filed for Chapter 11 Bankruptcy protection.

**January 28, 2009** – Heller Ehrman requested permission from the U.S. Bankruptcy Court to purchase a new $10.2 million malpractice insurance policy before their existing policy expired. Judge Dennis Montali denied this request, a ruling contrary to standard California practice.

*Without malpractice insurance, Iris's $100 million claims, if allowed, threatened to consume most or all of Heller Ehrman's remaining assets. Judge Montali had already approved a distribution plan for many unsecured creditors.*

*Law.com quoted Thomas Willoughby, representing Heller's creditors committee, saying, "We could look stupid buying it, or we could look stupid not buying it."*

**April 27, 2009** – The U.S. Bankruptcy Court set the claim bar date for April 27, 2009, for creditors to file claims.

**July 31, 2009** – Even after the claim bar date and Heller's bankruptcy filing, the firm sent a letter to Mr. Chin concealing the USPTO letters it had received in 2008. This letter reiterated assurances that "we will forward all correspondence and faxes which we may receive concerning your application from the U.S. Patent and Trademark Office." As a result, Mr. Chin continued to trust Heller Ehrman and remained unaware of their malpractice.

**July 1, 2010** – Heller Ehrman filed a list of unsecured creditors with the bankruptcy court on which Iris Biotechnologies appeared twice as a creditor. However, Iris was never notified of this critical filing, effectively leaving them unaware of their status and the proceedings that directly impacted their claims.

**August 22, 2011** – John Jeffrey of Dennison Associates in Canada informed Iris of Heller Ehrman's malpractices for the first time. Upon learning this, Iris immediately retained the prominent law firm Arnold & Porter LLP and instructed Dr. James A. Fox who had been Iris's trusted patent attorney at Heller Ehrman for seven years to file a petition with the USPTO to issue a new patent to Iris. Recognizing the urgency, Iris promptly filed malpractice claims within the one-year statute of limitations upon discovery of the evidence, thereby preserving their legal rights.

*(Because Heller Ehrman was already under Chapter 11 bankruptcy protection, the malpractice lawsuit was filed in bankruptcy court.)*

**August 20, 2012** – The patent at the heart of the dispute, *Artificial Intelligence System for Genetic Analysis*, held immense commercial potential worth billions of dollars when fully realized. Iris Biotechnologies asserted a $100 million claim against Heller Ehrman and filed a motion in the U.S. Bankruptcy Court in San Francisco to allow a late claim filing post-bar date, specifically targeting Heller Ehrman's malpractices. Thanks to the continued efforts of patent attorney Dr. James Fox, the patent was ultimately granted to Iris in May 2014, vindicating their claim to this critical intellectual property.

**November 20, 2012** – U.S. Bankruptcy Court Judge Dennis Montali officially recorded his ruling, confirming that Heller Ehrman had indeed failed to forward critical USPTO documents to Iris in March and October of 2008. Judge Montali stated:

*"Mr. Chin strikes me as a very intelligent, very successful, capable executive of a company that obviously is the result of much of his intelligence and learning. He couldn't have been warned many more times about the need to protect his own company's interests from the correspondence, to the delivery of the files, to the acknowledgment that the patent files were there for him and his company. Most importantly, his recognition that Heller was out of business and his view that Heller lacked competence not incompetent, but lacked competence. To me, that underscores the fact that the delay in acting and learning the fate of his company's filings with the Patent Office were things that he or Iris's counsel, if there were substitute counsel, could have ascertained."*

61

However, Judge Montali's ruling exhibited a marked bias by focusing solely on the first sentence of Mr. Chin's December 20, 2007 email to Dr. Fox: "It appears that Heller Ehrman has evolved into a firm that has problems not only with greed but also competence and ethics," while deliberately ignoring the immediately following sentence: "What happened has nothing to do with you, and I remain respectful of you, and I consider you a friend." This omission significantly prejudiced the ruling.

The ruling failed to consider the full context and all the clear evidence demonstrating why Iris could not have reasonably filed their claim before the April 27, 2009 bar date. For seven years, Dr. Fox successfully prosecuted all of Iris's patent applications and explicitly advised Iris to "do nothing" until hearing directly from the USPTO. After the bar date, Heller's July 31, 2009 letter deceptively reassured Iris that no office action had been received from the USPTO, which led Iris to believe there was no need for follow-up.

Critically, the ruling ignored the fact that Heller Ehrman was still Iris's legal patent representative, as the USPTO had denied Heller's petition to disengage. Dr. Fox remained the USPTO attorney of record for Iris until September 2011, underscoring that Heller had a continuing duty to inform Iris of correspondence. The court's failure to recognize this duty and the consequences of Heller's concealment resulted in a grossly prejudiced decision.

Additionally, Judge Montali's close professional ties to lawyers affiliated with Heller Ehrman and its counsel raised legitimate concerns about potential conflicts of interest affecting the impartiality of the ruling.

Judge Montali continued:

*"So I come to the conclusion, therefore, that as hard a result as it is for Iris, the factor of the reason for the delay and who has control over that delay lands squarely on Iris's lap and could not extricate it from the problem that it could have alleviated or mitigated on its own. Which is another way of saying that had Iris attended to its own protection of its own interest and come forth much more quickly, even after the claims bar date and asked to file a late claim, then indeed those factors would weigh less heavily against it and in favor of Heller. With the length of time of over two years of it not doing what it was cautioned to do and could have done and easily could have done, then I cannot extricate it from the consequences. So that is a long way of concluding that I am going to find that the Pincay and Pioneer factor weighed against Iris and in favor of Heller, and we will deny the motion to file a late claim. I mean, the claim has been filed, but by denying the motion, I am disallowing the claim as late filed claim."*

Immediately after delivering his ruling against Iris, Judge Montali physically choked for several seconds and then stated, "I can't speak." Notably, he refused to make eye contact throughout the court proceedings. This choking episode, captured in the official court audio record, is believed by Iris's representatives to be a physical manifestation of the judge's inner conflict over knowingly issuing an unjust ruling.

When Judge Montali denied Iris Biotechnologies the right to a jury trial on their $100 million claim against Heller Ehrman, he knowingly upheld a decision that effectively shielded Heller from financial liability. This was especially significant because Judge Montali had previously denied Heller Ehrman's request to spend $10.2 million on malpractice insurance that could have been used to compensate Iris for the egregious malpractices they suffered.

Judge Montali's ruling was not only inequitable but also illogical and contrary to established U.S. patent law. It appeared to cover up his prior error in disallowing Heller's expenditure on malpractice insurance. Despite clear evidence that Heller owed Iris money, Montali asserted that Mr. Chin could have hired a patent attorney to discover Heller's malpractices before or soon after the bar date. This ignores the reality that there was no reasonable or financially sound basis to suspect malpractice or to hire counsel solely for investigation purposes prior to the April 27, 2009 bar date, especially given that the USPTO had imposed no deadline for the patent application.

In 2010 and 2011, Iris continued to follow the instructions of their patent attorney, Dr. Fox. At no time were they informed by either Dr. Fox or Heller Ehrman about the departure of all patent attorneys from Heller's Menlo Park office, nor was Iris ever notified that Heller would stop forwarding critical mail. Iris prosecuted all of its patent applications worldwide diligently and on time, with the sole exception of the delays caused by Heller's misconduct.

Contrary to the ruling, the responsibility for the delay and control over it rests squarely with Heller Ehrman. Their malpractice, concealment of that malpractice, false reassurances, and deliberate misdirection caused the delay in securing Iris's foundational U.S. patent in the most commercially important market.

Heller's deceptive actions also delayed Iris's discovery of malpractice and hindered their ability to file a timely malpractice claim while insurance coverage was still in place. The ruling completely disregarded Mr. Chin's uncontested testimony that he had no issues with Dr. Fox's patent work and that he followed all of Fox's instructions.

Given all the evidence, it is inequitable to hold Iris responsible for the failure to file before the bar date, and irresponsible not to hold Heller accountable for their malpractice and concealment that caused the delay. Accordingly, the ruling should be overturned and Iris's claim allowed.

**December 5, 2012** – Based on the compelling evidence of Heller's actions preventing an earlier claim filing, Iris BioTech filed a notice of appeal with the United States District Court in San Francisco. However, Iris was denied even the opportunity for an oral hearing. (Iris also appealed to the Ninth Circuit Court, but again was denied an oral hearing.)

**Note:** As a direct result of Heller Ehrman's multiple malpractices against Iris Biotechnologies and their prolonged concealment compounded by Bankruptcy Judge Dennis Montali's evidently biased ruling that failed to hold Heller accountable, hundreds of Iris investors and their families endured significant hardship.

Despite Iris's rightful claim, they received no compensation from Heller Ehrman. The malpractice and ensuing legal battles ultimately contributed to a loss of approximately $20 million in my personal brokerage accounts a sum exceeding the average wealth of the U.S. top 1%.

Judge Montali's grossly unjust ruling inflicted severe financial, mental, and emotional harm on me. As a plaintiff in the United States legal system, I never imagined losing everything I worked for due to an unfair court decision.

From this experience, I learned that while people can take your money, no one not even yourself should ever be allowed to steal your time, peace, happiness, or love. With God as my witness, I hold firm belief that there will be a day of judgment for us all.

According to a Gallup Poll released on December 17, 2024, "Americans' confidence in their nation's judicial system and courts dropped to a record-low 35% in 2024. Public trust in the national government also plummeted to 26%."

# Chapter 8
# Cultivating Resilience and
# Learning from Failure

The entrepreneurial path is inherently strewn with obstacles, setbacks, and outright failures. The odds of success are daunting, and the personal and professional sacrifices required can be immense. Yet, it is precisely in facing these challenges head-on that entrepreneurs grow and thrive. In the face of such adversity, cultivating resilience and a growth mindset is not just helpful but absolutely essential.

One of our darkest chapters was the legal battle with our former law firm, Heller Ehrman, over the mishandling of our patent application. Their negligence cost us years of protection for our core AI technology, resulting in millions of dollars in lost value, legal fees, and opportunities that could never be recovered. The sense of betrayal and injustice was gut-wrenching, and the distraction of the lawsuit drained both our resources and morale.

But even amidst that pain, these experiences became unexpected teachers. They forced us to develop new reserves of strength, creativity, and strategic thinking. From each setback, we gleaned crucial insights that made us smarter, tougher, and more adaptable. We learned to anticipate and mitigate risks more effectively, to make difficult decisions with greater speed and confidence, and to rally together with a shared sense of purpose when the odds seemed stacked against us.

Critically, we also learned to shift our mindset and reframe failures as opportunities for growth and learning, rather than as endpoints or signs of defeat. Rather than dwelling

on what went wrong, we focused relentlessly on what we
could do better next time. We celebrated the effort,
ingenuity, and courage that went into noble attempts even
when they didn't succeed. And we became increasingly
comfortable with pivoting, iterating, and course-correcting
based on new information and changing circumstances.

For aspiring founders and leaders, my strongest advice is to
embrace adversity as an inevitable and even necessary part
of the growth process. Expect that things will go wrong,
you will make mistakes, and you will face criticism,
rejection, and setbacks. This is not a sign of failure but an
essential rite of passage on the road to success. But also
know that these very challenges are what forge great
entrepreneurs and build great companies.

Surround yourself with people who believe deeply in you
and your vision but who will also challenge you with
honest feedback and hold you accountable. Develop habits
and routines that keep you physically strong, mentally
sharp, and emotionally resilient. Prioritize your well-being
as fiercely as your business goals because your endurance
depends on it. And cultivate a deep, unwavering sense of
purpose your "why" that can sustain you through even the
toughest times.

Remember, failure is not the opposite of success but a
stepping stone to it. The only true failure is the failure to
learn, to grow, and to keep pushing forward. By reframing
setbacks as opportunities, maintaining laser focus on your
north star, and refusing to give up on yourself or your team,
you will build the resilience necessary to weather any storm
and emerge stronger, wiser, and more capable on the other
side.

In the biotech industry, resilience takes on added dimensions due to the unique challenges of the field:

1. Scientific setbacks: In biotech, years of work can be undone by a single failed experiment or clinical trial. It's crucial to develop the resilience to absorb these setbacks, learn from them, and pivot when necessary. Remember that even "failed" experiments often yield valuable data that can inform future directions.
2. Regulatory hurdles: Navigating the complex regulatory landscape in healthcare can be frustrating and time-consuming. Resilience here means maintaining patience and persistence while also being agile enough to adapt your strategy based on regulatory feedback.
3. Funding challenges: The capital-intensive nature of biotech means that funding challenges are common. Resilience in this context involves maintaining your vision and enthusiasm even when facing rejection from investors and being creative in finding alternative funding sources.
4. Ethical dilemmas: Biotech often involves grappling with complex ethical questions. Resilience here means staying true to your values even when faced with difficult decisions or external pressures.
5. Public scrutiny: As your work has direct implications for human health, you may face intense public and media scrutiny. Developing the resilience to handle this pressure while maintaining transparency and integrity is crucial.
6. Team morale: In the face of setbacks, maintaining team morale is critical. Resilient leaders find ways to keep their teams motivated and focused on the long-term mission, even during tough times.

7. Personal well-being: The high-stakes, high-stress nature of biotech entrepreneurship can take a toll on personal well-being. Building resilience means prioritizing self-care and maintaining a healthy work-life balance.
8. Technological obsolescence: The rapid pace of technological advancement in biotech means that your approach or technology could become obsolete. Resilience here involves staying adaptable and open to pivoting your strategy when necessary.
9. Competitive pressures: The biotech field is highly competitive. Resilience means staying focused on your unique value proposition and not getting derailed by every move your competitors make.
10. Long timelines: The extended timelines in biotech development can test anyone's patience and resolve. Resilience here means maintaining your commitment and enthusiasm over the long haul and celebrating small wins along the way.
11. Interdisciplinary challenges: Biotech often requires integrating knowledge from multiple scientific disciplines. Resilience in this context means being willing to continuously learn and adapt as you encounter new areas of knowledge.
12. Balancing science and business: Many biotech entrepreneurs come from scientific backgrounds and must learn to balance scientific rigor with business pragmatism. Resilience here involves being open to developing new skills and perspectives.
13. Managing uncertainty: Much of biotech involves working at the frontiers of scientific knowledge, which inherently involves high levels of uncertainty. Resilience means getting comfortable with this uncertainty and learning to make decisions with imperfect information.

14. Handling success: Interestingly, handling success can also require resilience. Rapid growth or sudden breakthroughs can bring their own set of challenges. Resilient leaders are able to navigate these transitions while maintaining their core values and vision.

15. Societal impact: The potential for significant societal impact in biotech can be both motivating and pressuring. Resilience here means staying connected to your core purpose while managing the weight of these expectations.

Cultivating resilience is not about becoming hardened, indifferent, or insensitive to setbacks. Rather, it's about developing the mental and emotional flexibility to absorb shocks, learn from them, and keep moving forward with renewed strength. It's about anchoring yourself firmly to your core purpose and values, even as you adapt your strategies, tactics, and expectations.

In biotech, perhaps more than in any other field, resilience is deeply tied to a sense of higher purpose. The potential to improve human health and save lives is an incredibly powerful motivator, fueling determination and perseverance through even the toughest and most prolonged challenges. Keeping this bigger picture in mind the patients whose lives you might change, the scientific frontiers you're pushing can become a wellspring of resilience that sustains you and your team.

At the same time, it's important to balance this sense of mission with realistic expectations and generous self-compassion. Not every experiment will succeed, and not every product will make it to market and that's okay. The key is to extract value from every experience, to analyze failures honestly without judgment, and to view each

71

setback as an opportunity for growth and learning rather than a permanent defeat.

Building a culture of resilience within your organization is equally crucial. This means creating a psychologically safe environment where people feel empowered to take calculated risks without fear of harsh criticism or punishment. It means cultivating a mindset where failure is viewed as an essential learning step, rather than a cause for blame or shame. It means fostering open communication, mutual support, and collective problem-solving. And it means celebrating not just the wins but also the effort, creativity, and courage that go into every attempt, regardless of outcome.

Remember, resilience is not a fixed trait or innate quality it's a skill that can be intentionally developed and strengthened over time. By consciously cultivating resilience in yourself and your team, you will be better equipped to navigate the inevitable ups and downs of the biotech entrepreneurship journey with grace and persistence. You'll be able to maintain your vision, enthusiasm, and emotional well-being even in the face of setbacks and prolonged uncertainty. Ultimately, this resilience will increase your chances of achieving breakthrough innovations that can transform human health.

In the end, the most successful biotech entrepreneurs are not those who avoid failure at all costs, but those who learn to fail forward, that is, to extract valuable lessons from every setback and use those lessons as fuel for future success. By embracing this mindset of resilience, continuous learning, and adaptive growth, you position yourself and your venture to make a lasting and meaningful impact in the challenging yet immensely rewarding world of biotechnology.

# Chapter 9
# Staying True to Your
# Integrity and Values

In the high-stakes world of entrepreneurship, where the pressure to succeed can be all-consuming, it can be all too easy to cut corners, compromise values, or make decisions that prioritize short-term gains over long-term integrity. Yet, in my experience, staying true to your moral compass is not just the right thing to do it is also the smartest strategy for the sustainable health and lasting success of your business.

Invariably, when we listened to our moral intuition and made the principled choice, even when it was difficult or costly in the moment, it paid off in the long run. We attracted team members, investors, and partners who shared our values and were committed to the right reasons not just profit. We built trust with regulators, customers, and advocacy groups who knew we could be counted on to do the right thing, even when no one was watching. And perhaps most importantly, we were able to look at ourselves in the mirror and feel proud of how we conducted ourselves and represented our industry.

On the flip side, we witnessed countless cautionary tales of companies that chased growth and profits at the expense of ethics only to pay a steep and sometimes irreversible price down the road. In biotech, cutting corners on safety, privacy, or informed consent is not just ethically wrong; it can be catastrophic for patients and utterly ruin public trust. As the saying goes, "You can't put a price on a clear conscience" and the long-term cost of losing one's ethical footing is far greater than any short-term gain.

# Chapter 9: Staying True to Your Integrity and Values

However, living up to one's principles is rarely black and white. It requires constant vigilance, humility, and course correction. As Iris grew and the stakes got higher, we had to work harder to stay true to our original vision and values. We had to be deliberate and transparent about defining and communicating our ethical standards. We also implemented systems and processes to hold ourselves accountable and ensure we were truly walking the talk not just paying lip service.

None of this was easy, and we certainly did not get it right all the time. But by making ethical integrity a non-negotiable core pillar of our culture and decision-making, we avoided many of the pitfalls that sunk other promising startups. More importantly, we built a company we could be proud of - one that made a positive, tangible difference in the world.

For other founders and leaders, my advice is to make your values and ethical principles explicit from day one. Know what you stand for and what lines you will not cross then embed these standards into every aspect of your business. Communicate them clearly and consistently to your team, your stakeholders, and yourself. And most importantly, live these principles out loud in your own actions and choices, setting the tone from the top.

Understand that there will be times when sticking to your values requires hard trade-offs and sacrifices. You may lose a lucrative deal, fire a high-performing but toxic employee, or admit to mistakes that bruise your ego. But you will gain something far more valuable in return: a reputation for integrity, self-respect that can weather any setback, and a purpose-driven business capable of standing the test of time.

Entrepreneurship is ultimately about stewardship the responsible management of resources, relationships, and the public trust in service of creating something meaningful and enduring. By anchoring that stewardship in an unwavering commitment to your values, you will not only sleep better at night you will also unleash the full potential of your venture to drive positive, lasting change. And what could be more rewarding than that?

In the biotech industry, the imperative to maintain integrity and adhere to strong ethical principles is particularly acute. Here's why:

1. Human impact: Biotech innovations often have direct implications for human health and well-being. The potential to help or harm is immense, making ethical considerations paramount.
2. Scientific integrity: The credibility of your research and products depends on maintaining the highest standards of scientific integrity. Any compromise here can have far-reaching consequences.
3. Regulatory compliance: The biotech industry is heavily regulated for good reason. Strict adherence to regulatory requirements is not just a legal necessity but also an ethical obligation.
4. Patient trust: Patients often put their lives in the hands of biotech innovations. Maintaining their trust through ethical practices is crucial.
5. Data privacy: With the increasing use of genetic and personal health data, protecting individual privacy is a critical ethical consideration.
6. Equitable access: There's an ethical imperative to ensure that life-saving innovations are accessible to those who need them, not just those who can afford them.

7. Animal welfare: For companies involved in preclinical research, ensuring ethical treatment of animal subjects is crucial.
8. Environmental responsibility: Biotech processes can have significant environmental impacts. Ethical companies strive to minimize these and contribute to sustainability.
9. Informed consent: In clinical trials and genetic studies, ensuring truly informed consent from participants is an ethical imperative.
10. Conflict of interest management: Given the complex web of relationships in biotech between researchers, companies, and healthcare providers, managing conflicts of interest transparently is essential.
11. Ethical use of technology: As biotech increasingly intersects with fields like AI and gene editing, there are new ethical frontiers to navigate.
12. Global health equity: Biotech companies have an opportunity to address global health disparities, raising ethical questions about resource allocation and research priorities.
13. Transparency in reporting: Both in scientific publications and in communications with investors and the public, maintaining transparency about results (both positive and negative) is crucial.

To navigate these ethical challenges in biotech, consider the following strategies:

1. Establish an ethics committee: Create a dedicated group within your organization to address ethical issues as they arise and to proactively develop ethical guidelines.
2. Implement robust compliance programs: Go beyond the minimum regulatory requirements to ensure that

ethical considerations are baked into every aspect of your operations.

3. Foster a culture of ethical awareness: Regularly discuss ethical considerations in team meetings and decision-making processes. Make it clear that raising ethical concerns is not just accepted but encouraged.

4. Engage with bioethicists: Consider bringing in outside experts to provide perspective on complex ethical issues.

5. Prioritize diversity and inclusion: Ensure that diverse voices and perspectives are included in your decision-making processes, especially when it comes to ethical considerations.

6. Develop clear guidelines for emerging technologies: As you work with cutting-edge technologies, proactively develop ethical guidelines for their use.

7. Engage with patient advocacy groups: These organizations can provide valuable perspectives on the ethical implications of your work from the patient's point of view.

8. Invest in ethics training: Provide ongoing ethics training for all employees, not just those in leadership positions.

9. Create accountability mechanisms: Develop systems to hold the organization accountable to its ethical standards, including consequences for ethical breaches.

10. Practice radical transparency: Be open about your ethical decision-making processes, both internally and externally.

11. Collaborate on industry-wide ethical standards: Work with other companies and organizations in your field to develop and promote ethical best practices.

12. Consider long-term impacts: When making decisions, consider not just immediate outcomes but potential long-term ethical implications.

By prioritizing integrity and ethical conduct, biotech companies can build deep, lasting trust with patients, healthcare providers, regulators, and the public. This trust is invaluable not only from a moral standpoint but also from a strategic business perspective. It can lead to stronger partnerships, smoother and faster regulatory approvals, improved talent acquisition and retention, and ultimately, greater long-term sustainability and success in a highly competitive industry.

Moreover, by maintaining high ethical standards, biotech entrepreneurs have a unique opportunity to shape the future of healthcare and biotechnology in profoundly positive ways. You're not simply building a company; you are contributing to the establishment of ethical frameworks and cultural norms that will guide the industry's growth and responsibility for decades to come. This leadership can help prevent abuses, protect vulnerable populations, and foster innovation that benefits society at large.

Remember that ethical leadership starts at the very top. As a founder or leader, your actions and decisions set the tone for the entire organization. When faced with difficult decisions, ask yourself not just, "Is this legal?" but more importantly, "Is this right?" Consider the broad impact of your choices on all stakeholders - patients, employees, investors, the scientific community, and society as a whole.

It is also important to recognize that ethical dilemmas in biotech are often complex, multifaceted, and nuanced. There may not always be clear-cut right or wrong answers. The key is to approach these issues thoughtfully, engage with diverse perspectives, including ethicists, patient advocates, and regulatory experts, and make decisions that you can stand behind with confidence and transparency.

Staying true to your integrity and values does not mean being rigid or unable to adapt to changing circumstances. Rather, it means having a clear and consistent moral compass that guides your decision-making, even as the specific application of those principles may evolve over time in response to new information, technologies, and societal expectations.

As you navigate your entrepreneurial journey in biotech, let your values be your compass and your guide. Use them as a framework for thoughtful decision-making, a foundation for building a resilient and positive organizational culture, and a beacon to attract like-minded partners and team members who will support and strengthen your mission. By doing so, you not only build a more sustainable and successful business, but you also actively contribute to shaping a more ethical, responsible, and respected biotech industry as a whole.

In conclusion, integrity in biotech entrepreneurship is not a constraint on success. It is a critical catalyst for it. By steadfastly staying true to your values, you create the foundation for lasting impact and meaningful innovation. You build a legacy not just of scientific and commercial achievement but also of principled, ethical leadership. And ultimately, you fulfill the true promise of biotechnology: to improve human health and well-being in ways that honor and uplift our shared humanity.

# Chapter 10
# Building the Future Through Collaboration

As I reflect on my entrepreneurial journey and the evolution of the biotech industry, one theme emerges again and again: the transformative power of collaboration. The challenges we face as a society whether in health, the environment, or social justice are far too complex, interconnected, and urgent for any one person or organization to tackle alone. Real, lasting progress demands that we break down silos and work together across disciplines, sectors, and geographies. Only through such collective effort can we hope to make meaningful and sustained impact.

At Iris Biotechnologies, collaboration was baked into our DNA from the very start. We recognized early on that to realize the full promise of precision medicine, isolated innovation was not enough. We needed to partner closely with academic researchers, clinical centers, patient advocates, regulatory bodies, and even competitors. We made a deliberate, strategic decision to prioritize collaboration over competition wherever possible, grounded in the belief that a rising tide lifts all boats and that shared knowledge accelerates breakthroughs. This collaborative mindset enabled us to leverage diverse expertise, pool resources, and move faster toward our shared goals.

Looking ahead, I believe the imperative for collaboration in biotech and beyond will only grow stronger. The COVID-19 pandemic starkly underscored how deeply interconnected our fates are and how critical collective action is in confronting global threats. At the same time, the accelerating pace of technological change from AI-driven drug discovery to advanced genomics is creating new opportunities and complex challenges that will require unprecedented cooperation, coordination, and trust among stakeholders worldwide.

Furthermore, fostering a culture of open collaboration can help address systemic inequities by ensuring that innovations reach diverse populations and that diverse voices shape the future of biotechnology. As entrepreneurs and leaders, our challenge and responsibility is to build bridges, nurture partnerships, and champion a collaborative ecosystem where shared success is the true north. Some of the key areas where I see collaboration as essential include:

- Advancing precision medicine through data sharing and interoperability
- Addressing health disparities through community-based participatory research
- Combating antimicrobial resistance through public-private partnerships
- Developing ethical frameworks for the use of AI and big data sets in healthcare and wellness
- Accelerating the translation of research into clinical practice through consortia models forming legal entities to solicit bids and buy on behalf of those involved in the consortiums.
- Fostering open innovation and pre-competitive spaces for shared problem-solving

No one company or institution can drive these agendas alone. Achieving real, transformative change will require a collective effort spanning the entire biotech and healthcare ecosystem, as well as cross-sector partnerships with technology, government, philanthropy, and civil society. It will require a willingness to share knowledge, resources, and credit in service of a larger, common goal. It will also require a new kind of leadership one that prioritizes collaboration over competition, long-term impact over short-term gain, and the greater good over narrow self-interest.

For aspiring entrepreneurs and innovators, my advice is to make collaboration a core part of your strategy and culture from the outset. Map out the key players and stakeholders in your field, and proactively reach out to explore areas of common cause. Build relationships based on trust, transparency, respect, and mutual benefit. And be willing to share your own ideas and assets in the spirit of open innovation and collective progress.

At the same time, be discerning about the partnerships you pursue. Look for collaborators who share your values and vision, who bring complementary strengths and diverse perspectives to the table, and who are genuinely committed to the hard work of alignment and execution. Set clear expectations and metrics for success, and be willing to course-correct or walk away if the partnership is not living up to its potential or compromising your core principles.

Ultimately, the measure of success in collaboration is not just what you achieve together, but how you achieve it. By modeling the kind of leadership, integrity, and partnership you wish to see in the world, you can not only accelerate your own impact but also help shape a more collaborative, inclusive, and resilient ecosystem for all.

The biotech industry has a unique opportunity and responsibility to lead the way in this regard. By harnessing the power of science, technology, and human ingenuity in service of the greater good, we can help build a future in which every person has the chance to live a healthy, fulfilling life. And by working together across boundaries and silos, we can unleash the full potential of our collective intelligence, creativity, and compassion to solve the grand challenges of our time.

This is the kind of future worth fighting for and above all, worth collaborating for. It is a future in which innovation is not just a means to an end, but a way of being and working together. The lines between competition and cooperation, profit and purpose, and individual success and collective impact blur in service of a higher calling. This is a future in which the entrepreneurial spirit is harnessed not just for personal gain, but also for lasting positive change and the greater good.

This is the future I have devoted my career and my company to building. And it is the future I invite you to help create one partnership, one innovation, and one act of collaboration at a time. Together, we can redefine what is possible and build a better world for all. That is the power and the promise of collaborative entrepreneurship.

In the biotech industry, collaboration takes on even greater importance due to the complexity of the challenges we face and the interdisciplinary nature of the work. Here are some additional thoughts on how to foster effective collaboration in biotech:

1. Cross-disciplinary teams: Build teams that bring together diverse expertise - biologists, chemists, data scientists, clinicians, engineers, and business strategists. This diversity of perspective can lead to breakthrough innovations.
2. Academic-industry partnerships: Foster strong relationships with universities and research institutions. These partnerships can provide access

to cutting-edge research, specialized facilities, and a pipeline of talent.

3. Patient engagement: Collaborate closely with patient advocacy groups and individual patients. Their insights can be invaluable in shaping research priorities and designing more patient-centric solutions.

4. Regulatory collaboration: Work proactively with regulatory bodies. Collaborative approaches like the FDA's breakthrough therapy designation can help accelerate the development of promising treatments.

5. International cooperation: Global health challenges require global solutions. Look for opportunities to collaborate across borders, sharing knowledge and resources to tackle common problems.

6. Precompetitive collaborations: Identify areas where companies can work together on shared challenges without compromising their competitive positions. This might include developing common standards, sharing non-proprietary data, or jointly addressing industry-wide issues.

7. Open innovation platforms: Consider creating or participating in open innovation initiatives that allow a wider community to contribute ideas and solutions to specific challenges.

8. Collaborative funding models: Explore innovative funding approaches like public-private partnerships, consortia models, or collaborative grants that bring together multiple stakeholders to fund ambitious projects.

9. Data-sharing initiatives: Participate in responsible data-sharing efforts that can accelerate research while protecting patient privacy and intellectual property.

10. Cross-sector partnerships: Look beyond the biotech industry for collaborators. Partnerships with tech

companies, environmental organizations, or social enterprises can bring fresh perspectives and complementary capabilities.

11. Collaborative clinical trials: Consider innovative trial designs that allow for collaboration between multiple companies or institutions, potentially accelerating the development process and reducing costs.

12. Mentorship and knowledge sharing: Foster a culture of mentorship within your organization and the broader biotech community. Share your experiences and learnings to help nurture the next generation of innovators.

13. Ecosystem building: Invest in building and strengthening the broader biotech ecosystem in your region. A thriving ecosystem benefits all participants through shared resources, talent pools, and knowledge networks.

14. Ethical collaborations: Work with ethicists, policymakers, and community leaders to develop ethical frameworks for emerging technologies. Collaborative approaches to ethics can help build public trust and ensure responsible innovation.

15. Crisis response collaborations: As the COVID-19 pandemic demonstrated, the ability to rapidly form collaborations in response to global health crises is crucial. Build the relationships and systems that will allow for agile collaboration when urgent needs arise.

Remember, effective collaboration is not just about formal partnerships or agreements. At its core, it means fostering a mindset of openness, curiosity, and shared purpose. It's about creating an environment where ideas can flow freely, diverse perspectives are genuinely valued, and the collective goal takes precedence over individual egos. This mindset empowers teams to break down barriers, challenge assumptions, and spark innovation that none could achieve alone.

As a leader, you play a crucial role in setting the tone for collaboration. Leadership in collaboration requires vulnerability and humility, showing that it's okay not to have all the answers, and that collective intelligence is a strength. Lead by example: model the behaviors you want to see - be open to new ideas, give credit generously, and demonstrate a willingness to learn from others. Create incentives and recognition systems that reward collaborative efforts and shared successes, not just individual achievements, fostering a culture where contribution and cooperation are celebrated.

At the same time, be mindful of the challenges that can arise in collaborations. Clear communication, well-defined roles and responsibilities, and transparent decision-making processes are essential to keep everyone aligned. Be prepared to navigate differences in organizational cultures, priorities, and timelines. And always keep the shared goal at the forefront to help align efforts, build trust, and resolve conflicts constructively. Developing mechanisms for continuous feedback and conflict resolution is vital to maintaining momentum and ensuring all voices are heard.

Looking ahead, the biotech industry holds enormous potential to drive transformative change in human health, environmental sustainability, and beyond. However, realizing this potential will require us to think and work in fundamentally new ways. It will require us to break down silos, embrace transparency, and actively co-create solutions to our most urgent challenges. Furthermore, it demands a commitment to ethical considerations and social responsibility, ensuring innovations benefit society equitably.

By embracing collaboration as a core strategy and value, we can accelerate innovation, expand the impact of our work, and build a more resilient and inclusive biotech ecosystem. We can tackle bigger challenges, take on more ambitious projects, and create solutions that truly meet the complex and evolving needs of our world. Collaboration also opens doors to novel funding models and cross-sector partnerships that can amplify resources and expertise beyond traditional limits.

In doing so, we advance not only our individual enterprises but also contribute to a larger movement of positive change. We become part of something bigger than ourselves, a global community of innovators and problem-solvers working together to shape a better future. This collective approach is essential for addressing systemic issues that no single entity can solve, such as global pandemics, climate change, and health disparities.

So, as you build your venture or career in biotech, I encourage you to think broadly and strategically about collaboration. Seek out diverse partners, be generous with your knowledge and resources, and always look for opportunities to create shared value. By doing so, you'll not only increase your chances of success but also help build the kind of collaborative, innovative, and impactful biotech industry that our world so urgently needs. Remember, the strongest innovations emerge from ecosystems where trust, respect, and shared ambition thrive. Make those your foundation.

# Chapter 11
# Shaping a Purpose-Driven Legacy

As I enter the later chapters of my entrepreneurial journey, I find myself reflecting more and more on the question of legacy. What will I leave behind? How will my work and my life have mattered? What kind of world do I want to help create for future generations? These are questions that transcend business they strike at the heart of who we are, what we value, and how we hope to be remembered.

These are not just philosophical musings, but urgent and practical imperatives. In an era of existential challenges like climate change, inequality, and global health crises, the stakes for innovation and entrepreneurship have never been higher. The decisions we make and the actions we take today will reverberate for decades to come. We are no longer in a time where incremental progress is enough. We need bold, principled action that aligns long-term value creation with immediate, meaningful impact.

For me, the ultimate measure of success is not just financial or even scientific, but moral and societal. It is about the lives we touch, the communities we strengthen, and the planet we protect. It is about creating shared prosperity and well-being and leaving the world better than we found it. True innovation must not only advance technology but uplift humanity. It must heal, not harm; connect, not divide; empower, not exploit.

This has been the North Star Guiding Iris Biotechnologies from its inception. Our vision has always been about more than just developing cutting-edge diagnostics and advancing precision medicine, as important as those goals are. It has been about harnessing the power of science and

88

technology to improve people's lives and address some of the most pressing challenges facing humanity. We recognized early on that biotechnology is not just a tool for treatment it's a platform for transformation, with the potential to redefine how we think about health, equity, and the future of care.

This purpose-driven approach has influenced every aspect of our business from the problems we choose to work on, to the partners we engage with, to the way we measure and report our impact. It has also shaped our culture and values, attracting team members and stakeholders who share our commitment to making a positive difference in the world. Purpose, for us, is not an afterthought or a corporate slogan. It is embedded in our operating model, our stakeholder relationships, and the DNA of our innovations.

As I look to the future, I am more convinced than ever that this kind of purpose-driven entrepreneurship will be essential to tackling the grand challenges of our time. We need a new generation of leaders and innovators who are not just technically skilled but also ethically grounded, who are driven not just by profit but by purpose, and who are committed to creating value for all stakeholders, not just shareholders. Leadership today requires moral clarity, cross-disciplinary fluency, and the courage to build systems that prioritize justice, access, and sustainability.

We need businesses and institutions that are designed for social and environmental impact, not just efficiency and growth. We need an economic system that rewards cooperation and shared prosperity, not just competition and individual gain. We need a global community that works together across borders and sectors to solve our common problems and create a better future for all.

In short, we need an economy of empathy powered by science, guided by ethics, and driven by collective ambition. The future we imagine is only possible if we build it together, piece by piece, across industries and generations.

This may sound idealistic, but I believe it is also deeply pragmatic. In a world of complex, interconnected challenges, the old models of business and leadership are no longer fit for purpose. The companies and leaders that will thrive in the coming decades will be those that embrace a new paradigm of purpose-driven, collaborative, and equitable value creation. This isn't just a shift in language it's a shift in how we define success, how we build trust, and how we align innovation with the needs of people and the planet.

For aspiring entrepreneurs and changemakers, my advice is to start with your why. What is the problem you are passionate about solving? What is the difference you want to make in the world? Let that purpose be your compass and your fuel for the journey ahead. Purpose is not a luxury it's a strategic necessity. It's what keeps you grounded during setbacks and focused during periods of growth.

Build a team and a culture that shares your values and your vision. Surround yourself with diverse perspectives and skills, and create an environment of trust, creativity, and continuous learning. Be willing to experiment, iterate, and adapt as you go, always keeping your ultimate impact in mind.

Remember: the strength of your team is not just in what they know, but in how they think, how they care, and how they challenge each other to be better. Culture is your long-term differentiator.

Cultivate a mindset of collaboration and partnership. Map out the ecosystem of stakeholders and allies who share your goals, and look for ways to leverage your unique strengths in service of the greater good. Be open to new business models and unexpected partnerships that can accelerate your impact. Success in today's world is rarely achieved in isolation. It is co-authored with communities, competitors, researchers, regulators, and visionaries across disciplines.

Most importantly, you should never lose sight of the people and communities you are serving. Listen to their needs and aspirations, and co-create solutions with them, not for them.

Measure your success not just in terms of financial metrics but also in terms of the real-world outcomes and experiences of those you seek to help. Empathy, humility, and proximity to the people you serve are not soft values they are hard advantages. They ensure your solutions are relevant, inclusive, and truly transformative.

This is the kind of entrepreneurship that I believe will define the 21st century entrepreneurship that is driven by purpose, grounded in values, and focused on creating lasting, positive change. It is entrepreneurship that sees beyond the bottom line, that dares to challenge convention, and that recognizes our shared responsibility to the planet and its people. It is the kind of entrepreneurship that has guided my own journey with Iris Biotechnologies, and that I hope will inspire and empower future generations of innovators and leaders.

As I reflect on my own legacy, I am filled with gratitude for the incredible people and experiences that have shaped my path. From the early days of bootstrapping and experimentation to the hard-won successes and painful setbacks, to the enduring relationships and shared triumphs, every step has been a lesson and a gift.

Each phase of the journey no matter how uncertain carved wisdom, resilience, and perspective into the core of who I am.

I am grateful to the mentors and role models who have guided and inspired me, to the team members and partners who have joined me on this journey, and to the patients and communities who have trusted us to make a difference in their lives. I am grateful to our investors for the opportunity to pursue my passions and to have contributed in some small way to the betterment of the world. But most of all, I am grateful for the chance to live and lead with purpose to build not just a business, but also a mission.

But I also know that my legacy is not mine alone to shape. It is a collective legacy, one that belongs to all of us who dare to dream big and work hard in service of a better future. It is a legacy that will be carried forward by the next generation of entrepreneurs and leaders, who will stand on our shoulders and take our work to new heights. Legacy is not just what we leave behind it is what we set in motion.

To them, I offer my heartfelt encouragement and my deepest respect. The road ahead will not be easy, but it will be worth it. The challenges you face will be daunting, but so too will be the opportunities to make a real and lasting difference. The setbacks you encounter will test your resolve, but they will also forge your strength and your character. You are the architects of the future, and every courageous step you take helps lay the foundation for a world worth inheriting.

Remember that you are part of a larger story a story of human progress and possibility that stretches back through the generations and forward into the future. Your role in that story is unique and precious, and it is yours to write with courage, compassion, and conviction. Do not underestimate the ripple effect of your integrity, your innovation, and your vision. Every act rooted in purpose adds to a legacy that transcends any single life or venture.

So dream boldly, act with integrity, and never give up on building the world you want to see. Embrace the power of purpose, the value of collaboration, and the imperative of equity. And know that you are not alone on this journey you are part of a global community of change-makers who share your vision and your values.

Together, you are not just launching companies you are shaping culture, shifting paradigms, and redefining what leadership looks like.

Together, we can shape a future in which every person has the chance to live a healthy, fulfilling life, in which every community can thrive and prosper, and in which every generation can look forward to a brighter tomorrow. That is the legacy we are called to build, and it is the legacy that will define our time on this earth.

Let it be a legacy rooted in justice, fueled by innovation, and remembered for its humanity.

As for me, I will continue to do my part to use my voice and my platform to champion the kind of entrepreneurship and leadership that I believe the world needs now more than ever. I will continue to learn and grow, to support and mentor, to advocate and agitate for change. And I will continue to hold fast to the belief that a better world is not only possible but also inevitable if we have the courage and the will to make it so. The work is far from over, and I remain as committed as ever to lifting others, sharing knowledge, and amplifying the mission that has guided me from the beginning.

That is my promise, my purpose, and my legacy. And it is one that I invite you to share and to build upon, in your own unique way, with your own unique gifts and passions. In the end, the true measure of our legacy will not be the monuments we build or the accolades we receive, but the lives we touch and the world we leave behind. Legacy, after all, is written not in stone but in the stories and futures of those we empower.

Let us go forth, then, with hearts full of hope and hands ready for service. Let us seize the opportunities before us and rise to the challenges of our time. Let us build a legacy of purpose, compassion, and progress that will endure long after we are gone. And let us never forget the power we hold, as entrepreneurs and as human beings, to shape a better future for all. This is our invitation to make meaning, to leave light, and to lead with love.

This is our calling, our responsibility, and our opportunity. May we answer it with courage, with compassion, and with unwavering commitment to the greater good! And may we always remember that the true reward of a life well-lived is not in the destination, but in the journey, and in the lives we are privileged to touch along the way. Because in the end, it's not about what we build it's about who we become, and how deeply we cared along the way.

In the context of biotech entrepreneurship, shaping a purpose-driven legacy takes on additional dimensions:

1. Scientific advancement: Your legacy includes the contributions you make to scientific knowledge. Even if a particular product doesn't reach the market, the research and discoveries made along the way can advance the field and pave the way for future breakthroughs.
2. Patient impact: In biotech, your legacy is measured in lives improved or saved. Keep sight of the real people who stand to benefit from your work.
3. Ethical leadership: By setting high standards for ethical conduct in biotech, you help shape the moral framework of an industry that has profound implications for human health and well-being.
4. Environmental stewardship: Consider how your work in biotech can contribute to environmental sustainability. From developing cleaner manufacturing processes to creating bio-based alternatives to petrochemicals, there are many ways biotech can positively impact the planet.
5. Mentorship and education: Your legacy includes the next generation of scientists and entrepreneurs you inspire and nurture. Consider how you can contribute to STEM education and provide opportunities for young people to engage with biotech.
6. Policy influence: As a biotech leader, you have the opportunity to shape policies that govern research, drug development, and healthcare. Use your voice and expertise to advocate for policies that advance science and expand access to care.
7. Global health equity: Consider how your work can address health disparities and contribute to improving healthcare access in underserved communities around the world.
8. Technological innovation: Your legacy may include new tools, platforms, or methodologies that change

how biotech research and development are conducted.

9. Interdisciplinary bridges: By fostering collaboration between biotech and other fields (like AI, nanotechnology, or environmental science), you can help create new paradigms for solving complex problems.

10. Cultural change: Through your leadership, you can contribute to changing the culture of the biotech industry - making it more diverse, inclusive, collaborative, and purpose-driven.

As you build your career and your company in biotech, regularly reflect on the broader impact of your work. Ask yourself not just "What are we achieving?" but "Why does it matter?" and "Who will benefit?" Let these questions guide your strategic decisions and your day-to-day actions. Let them become the compass that ensures you're not just moving fast but moving in the right direction.

Remember that in biotech, perhaps more than in any other field, your work has the potential to profoundly impact human lives and the future of our planet. This is both a great responsibility and an incredible opportunity. By staying true to your purpose, maintaining your integrity, and always striving to create value for society, you can build a legacy that extends far beyond financial success or scientific accolades.

You're not just developing products you're shaping the health, well-being, and sustainability of future generations. Your legacy in biotech is not just about what you achieve, but also about how you achieve it. It's about the culture you create, the ethical standards you uphold, and the way you treat people along the way. It's about using your knowledge and resources not just for personal gain, but also for the greater good.

Every hiring decision, every lab protocol, every public statement these all speak to the kind of legacy you're creating. As you move forward in your entrepreneurial journey, I encourage you to think expansively about your potential impact. How can your work in biotech contribute to solving some of the grand challenges facing humanity? How can you use your platform to advocate for positive change in the industry and beyond? How can you inspire and empower others to pursue purpose-driven innovation? And how can your success model a new kind of leadership one where compassion and collaboration drive progress just as much as competition and capital?

Building a purpose-driven legacy in biotech is not always easy. It may require making difficult choices, taking stands that aren't popular, or sacrificing short-term gains for long-term impact. But it is infinitely rewarding. It allows you to align your work with your deepest values and to know that your efforts are contributing to a better world.

In the moments when you're tested, remember: long-term trust is built through short-term courage. So as you build your company and pursue your goals, keep your eyes on the horizon. Think not just about the next quarter or the next year, but also about the impact your work will have decades from now.

Let your purpose be your guide, your integrity be your foundation, and your commitment to making a positive difference be the legacy you leave behind.

Success that endures is rooted in significance - let that be your measure.

In doing so, you'll not only build a successful career or company you'll contribute to shaping a biotech industry that is more ethical, more equitable, and more impactful.

You'll be part of writing a new chapter in the story of human progress, one where the power of science and technology is harnessed for the benefit of all. You'll help redefine what it means to lead in this space fusing innovation with empathy, and profits with purpose.

This is the true promise and potential of biotech entrepreneurship. It's a path that offers not just professional success, but the opportunity to make a meaningful difference in the world.

It's a chance to be part of something bigger than you and to contribute to solving some of the most pressing challenges of our time. Your journey can become a bridge connecting scientific excellence to human need.

As we conclude this exploration of entrepreneurship and legacy in biotech, I hope you feel inspired and empowered to pursue your own purpose-driven journey. Remember that every great achievement starts with a single step, every breakthrough begins with a question, and every positive change starts with someone who dares to imagine a better way. Dare to be that someone. Dare to ask more. Dare to give more.

You have within you the potential to create extraordinary value not just economic value, but human value. You have the power to improve lives, advance science, and contribute to a healthier, more sustainable world. That is the legacy you can build through purpose-driven entrepreneurship in biotech. And in doing so, you elevate not just your career but the entire field you touch.

So go forth with courage and conviction.

Pursue your purpose with passion and perseverance.

Build your legacy not just in the products you create or the profits you generate, but also in the lives you touch and the positive change you catalyze.

The world is waiting for your contribution, for your innovation, for your leadership.

**The future won't wait and neither should your vision.**

May your journey be bold, your impact profound, and your legacy be one that inspires generations to come.

The future of biotech and indeed, the future of our world depends on leaders like you who are willing to dream big, work hard, and never lose sight of the profound purpose that underlies this incredible field. **Lead with integrity. Create with purpose. Influence with humility.**

Here's to your journey, your impact, and the lasting legacy you will create.

The world is counting on you.

**Make it count. Never forget, the best breakthroughs begin in the heart before they reach the lab.**

# Section 2: Precision Medicine: Health and Wellness

# Chapter 12
# The Journey to Health and Wellness

On a crisp Friday in October 2007, the Saratoga Rotary Club hosted a guest speaker who would unknowingly set in motion a profound personal health journey. A young cancer survivor stood before the group, sharing her inspiring story of commitment and resilience. She spoke of her decision to run a marathon and raise money for cancer research after recovering from her own battle with the disease.

Among the audience sat a Rotary member, whose father was a prostate cancer survivor. As the young woman's words resonated through the room, a spark ignited within this listener – a spark that would soon burst into a flame of determination. In that moment, something shifted. The story wasn't just touching it felt like a call to action.

Inspired by the speaker's courage and driven by a desire to make a difference, the Rotary member made a life-changing decision that day. Despite never having run more than a few miles at a time, that member committed to completing a full marathon – 26.2 grueling miles – to raise funds for cancer patients. This wasn't just about personal achievement; it was about joining a larger fight against a disease that had touched so many lives.

It became a deeply personal mission, fueled by the hope of making even a small difference in someone else's battle. This was not a casual goal; it was a vow to step far outside of comfort zones for something greater than oneself.

The journey began with joining the Leukemia and Lymphoma Society's Team In Training program. This comprehensive training regimen was designed to transform ordinary individuals into marathon runners while raising funds for a worthy cause. The program offered more than just physical preparation; it provided a support system, expert guidance, and a sense of community among participants, all working toward the same goal.

There was accountability, shared struggle, and a contagious energy that lifted even the most doubtful runners. Participants weren't just training for a race they were becoming part of a movement.

On the first day of training, the magnitude of the challenge ahead became clear. Gathered at the local high school track, the groups of aspiring marathoners learned about the training process, met their coaches (including a former Olympian), and were introduced to the staff that would guide them through warm-ups, cool-downs, and the long journey ahead. The initial 2-mile combination of running and walking left many, including our protagonist, breathless. But it was just the beginning.

That day also served as a humbling reminder that transformation doesn't happen overnight it begins with small, committed steps. There was laughter, nervous chatter, and even doubt but also a sense of collective resolve.

Over the next four months, the training intensity steadily increased. Almost daily solo runs complemented regular evening sessions at Los Gatos High School. During the weekends, the group ran at locations ranging from Portola Valley to Santa Cruz. The regimen was demanding, requiring dedication, time, and physical effort. But with each passing week, endurance improved, and what once seemed impossible began to feel achievable.

Each mile was a milestone not just on the road, but also in mindset. Blisters, sore muscles, and early mornings became badges of honor, quietly affirming the commitment made on that October day. Through the sweat and setbacks, confidence took root, and purpose kept every footstep focused.

A month before the marathon, the training reached its peak. The schedule included running a half marathon almost every day for a week – a grueling test of endurance that pushed bodies and minds to their limits. This wasn't just about distance anymore it was about testing resolve. Two weeks before the big day, an 18-mile run was accomplished, instilling a sense of confidence and readiness. For the first time, crossing the finish line felt truly possible.

Finally, the day of the Marathon arrived. The atmosphere was electric with excitement and nervous energy. A sign at the check-in station boldly proclaimed, "It's going to happen!" – a simple yet powerful affirmation that brought a smile to many faces and calmed jittery nerves. For our runner, the goal was clear: complete the marathon in less than 5 hours. The crowd buzzed with energy - athletes, volunteers, supporters all united by purpose.

As the race began, adrenaline surged. Running among so many fit, determined individuals was exhilarating. The first half of the course flew by, completed in less than 2.5 hours. Buoyed by this strong start, there were plans to run even faster for the second half. Every stride felt strong, every breath steady a rhythm of triumph unfolding.

However, marathon running, like life itself, often doesn't go according to plan. An unexpected 15-minute wait to use a portable toilet proved to be a critical mistake. This delay led to a buildup of lactic acid, causing severe muscle fatigue and soreness. Upon resuming the run, both legs cramped up painfully, making it nearly impossible to continue. Momentum was lost, and in its place came the harsh reality of physical limits.

It was at this moment that a crucial lesson was learned – one that had been shared in a training session our runner had unfortunately missed. The importance of replenishing electrolytes during such a long and demanding physical endeavor became painfully clear. These essential minerals play a vital role in muscle function, affecting everything from hydration to nerve signals and muscle contractions. Neglecting this one piece of advice would come to define the next stretch of the race.

The remainder of the race became a battle of will against physical limitations. Cramps kept returning, necessitating frequent stops to massage aching legs. The continuous climb at mile 19 tested not just physical endurance but mental fortitude as well. Each step felt heavier, every incline steeper but quitting was never entertained.

Around mile 21 or 22, another challenge emerged. Severe blisters developed, with one popping so loudly it sounded like a gunshot – a startling and painful experience. Running became increasingly difficult, each step a testament to determination and grit. Shoes now soaked with sweat and pain, the trail blurred with discomfort, but the goal still glimmered ahead.

In this moment of struggle, the true spirit of the Team In Training program shone through. Staff members, noticing the difficulty, provided on-the-spot medical treatment, enabling the run to continue. This act of support embodied the collective spirit of the marathon – a shared journey where strangers become allies in the face of adversity. It wasn't just about one runner anymore it was about the mission, the team, and the cause.

The final miles were a blur of pain, fatigue, and sheer willpower. Quitting was not an option, not after months of training and with the finish line in sight. Even if it meant alternating between slow running and walking, forward progress continued. Pride mingled with pain, and every aching stride became a symbol of perseverance.

In a moment that felt almost cinematic, a coach who was a former Olympian appeared about a quarter mile from the finish line. Running alongside our struggling marathoner, this experienced athlete provided a final boost of motivation. Somehow, in the presence of this elite runner, the pain and exhaustion faded into the background.

With renewed energy, the last quarter mile was completed in stride with strength and determination. That final push powered not just by muscle, but also by spirit carried me home. I was the person that completed the 26.2-mile Napa Valley Marathon.

Crossing the finish line was more than just the end of a race; it was the culmination of a transformative journey. For someone born prematurely and plagued by frequent childhood illnesses, including a near-fatal bout of typhoid fever, completing a marathon held profound significance. It wasn't just about fitness it was about reclaiming a body that had once struggled to survive. In that moment, every setback, every hospital stay, every doubt was eclipsed by the finish line.

But the true victory extended beyond personal achievement. This journey raised vital funds for cancer patients, contributing to the larger fight against a disease that affects millions. It demonstrated the power of community, the importance of perseverance, and the impact that one-inspired individual can have when they commit to a cause greater than themselves. It proved that healing is not only possible but also contagious. One person's journey can spark another's hope.

The marathon experience provided valuable lessons that extended far beyond running. It highlighted the importance of proper preparation, the crucial role of support systems, and the unpredictable nature of any significant undertaking. It showed that setbacks and unexpected obstacles are not just challenges to be overcome, but opportunities for growth and learning. Each obstacle carried wisdom, and each mile taught resilience.

Most importantly, this journey served as a testament to the incredible potential within each of us. It proved that with determination, support, and a worthy goal, we are capable of achieving things we never thought possible. Whether it's running a marathon, overcoming illness, or making a difference in the lives of others, the human spirit has a remarkable capacity for growth, resilience, and triumph. It starts with a decision, is carried by discipline, and finishes with heart.

# Chapter 13
# Evolution of Cancer Care and Reproductive Health

The landscape of healthcare has undergone significant transformations over the past seven decades, particularly in the realms of cancer care and reproductive health. These changes reflect not only advancements in medical science but also shifts in societal attitudes and legal frameworks. Together, they paint a picture of progress shaped by science, policy, and evolving public consciousness.

In the fight against cancer, progress has been steady and encouraging. From 1950 to 1960, the death rate due to cancer in the United States stood at 193.9 per 100,000 populations. This rate continued to climb, reaching its peak in 1990 at 216 deaths per 100,000. However, since then, there has been a gradual but consistent decline. By 2019, the rate had fallen to 146.2 per 100,000 – a reduction of nearly 33% from the peak. This turning point marked a significant shift in the nation's ability to manage and treat a once-dreaded disease.

This improvement can be attributed to several factors. Advancements in cancer care have led to more effective treatments and better outcomes for patients. Early detection methods have improved, allowing for intervention at stages when cancer is more treatable.

Additionally, public health initiatives have had a significant impact, particularly in reducing smoking rates, which has played a crucial role in decreasing lung cancer incidence and mortality. Public education campaigns, tobacco regulations, and behavioral shifts all converged to reduce major cancer risk factors.

Research funding has been a cornerstone of these advancements. Increased investment in cancer research has led to breakthroughs in understanding the mechanisms of cancer development, identifying risk factors, and developing innovative treatment approaches. From targeted therapies to immunotherapy, these research efforts have opened new avenues for combating cancer and improving patient outcomes.

Federal and private sector funding spearheaded by efforts such as the National Cancer Act of 1971 laid the foundation for a modern era of oncology. Today, cancer care is increasingly personalized, with genetic profiling and precision medicine guiding treatment plans.

Parallel to the developments in cancer care, the field of reproductive health has seen its own revolution. The introduction of the birth control pill in 1960 marked a turning point in family planning and women's health. This innovation gave women unprecedented control over their reproductive choices, leading to significant societal changes. Access to reliable contraception empowered women to pursue higher education, careers, and long-term planning with greater autonomy.

The impact of this new contraceptive option quickly became apparent in abortion statistics. In 1960, before the pill was widely available, the U.S. abortion rate was approximately 0%. By 1973, it had climbed to about 3%. This increase reflects not only the growing availability of abortion services but also changing attitudes towards reproductive rights.

The landmark Roe v. Wade decision in 1973 further cemented the legal framework for reproductive autonomy, legitimizing access to safe abortion as a constitutional right until its reversal in 2022. These legal and cultural developments reveal the complex interplay between medicine, law, and personal agency in shaping reproductive health policy.

A watershed moment came in 1973 with the Roe v. Wade Supreme Court decision, which legalized abortion nationwide. Following this ruling, the abortion rate saw a sharp increase, reaching 10% by 1980. This rise can be attributed to increased access to legal abortion services and greater public awareness of reproductive rights.

According to the Guttmacher Institute, the number of abortions in the U.S. reached its peak in 1990, with 1.6 million procedures performed that year. This figure represents the highest point in a trend that had been building since the legalization of abortion. Since then, the total number of abortions has fluctuated but generally declined, influenced by changing laws, improved contraception, and evolving public opinion.

To put these numbers in a global context, the World Health Organization reports that approximately 73 million induced abortions occur worldwide each year. This accounts for a significant proportion of pregnancies, 61% of all unintended pregnancies and 29% of all pregnancies overall end in abortion. These statistics underscore that abortion is not only a national issue but also a global one, revealing the widespread need for access to reproductive health services and preventive care.

More recent data from the Centers for Disease Control and Prevention (CDC) provides a snapshot of current trends in the United States. In 2021, there were 11.6 abortions per 1,000 women ages 15 to 44. It's important to note that this figure excludes data from several states, including

California, the District of Columbia, Maryland, New Hampshire, and New Jersey. This rate represents a significant decrease from 1980, when the CDC reported 25 abortions per 1,000 women ages 15 to 44 across all 50 states and D.C. This decline suggests that while abortion remains a common procedure, changes in access to contraception, sex education, and policy have likely influenced overall trends.

The landscape of abortion care has also evolved with the introduction of medication abortions. In 2023, there were about 642,700 medication abortions in the United States, accounting for nearly two-thirds of all abortions nationwide. This marks a significant shift from 2000, when medication abortions were not available.

This transformation has made abortion care more private, earlier in gestation, and depending on legal status more accessible for many. It has also intensified debates about regulation, safety, and availability.

Despite legal changes and varying access across states, recent data suggests that abortion rates have remained relatively stable over the long term. According to the Guttmacher Institute, there were 16.3 abortions in the U.S. per 1,000 women ages 15 to 44 in 1973. In 2023, this rate was 15.6, indicating that while the number of abortions has fluctuated over the years, the overall rate has not changed dramatically since legalization. This consistency raises important questions about the root causes of unintended pregnancies and the effectiveness of preventive strategies.

The most recent findings from the Monthly Abortion Provision Study provide insight into the current state of abortion in the U.S. In 2023, the first full calendar year after the U.S. Supreme Court's decision in *Dobbs v.*

*Jackson Women's Health Organization* overturned *Roe v. Wade*, an estimated 1,026,690 abortions occurred in the formal healthcare system. This represents a rate of 15.7 abortions per 1,000 women of reproductive age marking a 10% increase since 2020, the last year for which comprehensive estimates were available.

Notably, this is the highest number and rate measured in the United States in over a decade, reflecting not only demand but also the increasing use of telemedicine and out-of-state care in response to changing laws.

These statistics underscore the complex and evolving nature of reproductive healthcare in the United States. They reflect changes in societal attitudes, legal frameworks, and medical technologies. They also reinforce the critical need for accessible, evidence-based reproductive healthcare including contraception, family planning education, and safe, legal abortion options.

While access remains essential, so does prevention. A global total of 73 million induced abortions per year signals an urgent need for improved strategies to reduce unintended pregnancies. Comprehensive education, affordable contraception, and public awareness are key. The most effective time to discuss abortion is not when a woman is already pregnant, but long before through proactive support and information.

Some women advocate that abortion is about their bodies and their rights. However, what they are aborting is not their bodies but the unborn. This perspective invites a deeper ethical conversation about bodily autonomy, fetal development, and the moral responsibilities of individuals and society. Balancing compassion, rights, and responsibility remains at the heart of the debate.

The evolution of cancer care and reproductive health over the past seven decades illustrates the dynamic nature of healthcare. Advancements in medical science have led to improved outcomes and increased options for patients. At the same time, shifts in societal attitudes and legal frameworks have reshaped the landscape of reproductive rights and access to care. These changes remind us that healthcare is not only a clinical field but also a human one shaped by values, struggles, and collective decisions.

As we move forward, it's clear that both cancer care and reproductive health will continue to be areas of intense focus, research, and debate. The goal remains to improve health outcomes, provide comprehensive care, and ensure that individuals have access to the information and services they need to make informed decisions about their health, well-being, and the unborn. Respect for life, empathy for those facing difficult choices, and a commitment to prevention should guide future policy and healthcare innovations.

You were once a vulnerable unborn baby. Be grateful to all who cared for you and let that gratitude inspire compassion, responsibility, and respect for life in all its stages.

# Chapter 14
# The Dawn of Precision Medicine

The advent of precision medicine marks a transformative new era in healthcare, promising to fundamentally change how we understand, prevent, and treat diseases on an individual level. This innovative approach tailors medical treatment to the unique genetic, environmental, and lifestyle characteristics of each patient, positioning itself to revolutionize the medical landscape in profound ways.

Before the COVID-19 pandemic, companies like IRIS Wellness Labs were already at the forefront of this revolution, pioneering individualized and comprehensive scientific analyses that far exceeded the capabilities of traditional medicine. While conventional health and wellness assessments often provided limited, generalized information, IRIS aimed to empower physicians and patients alike by delivering deep, personalized data to optimize medical treatment and overall well-being.

After years of meticulous preparation and technological advancement, Iris Biotechnologies found itself at a watershed moment. The company was poised to take a leading role in the precision medicine movement, capitalizing on the dramatic decrease in DNA sequencing costs over the past two decades. This cost reduction—dropping from billions of dollars to just a few hundred dollars per genome—has opened the door to unprecedented access and scalability in genetic analysis.

The foundation for this revolution was laid by the Human Genome Project, which has fundamentally accelerated our understanding of human biology and medicine. However, the insights gained from sequencing the human genome revealed that the genome sequence alone tells only part of the story. Equally critical to health are factors such as gene expression patterns, epigenetic modifications, and the complex interactions within the microbiome—each adding crucial layers of biological complexity.

Recent studies have shed new light on the dynamic evolution of the human genome. Over the past 5,000 years, our genetic makeup has undergone prolific changes, paralleling rapid population growth and an explosion of genetic mutations. These evolutionary shifts carry profound implications for how we understand susceptibility to diseases, adaptation to environments, and responses to treatments.

Deep sequencing research has identified and chronologically mapped over a million single nucleotide variants (SNVs)—specific points in the DNA where a single letter differs from the typical sequence. Analysis of the genomes of approximately 6,500 African and European Americans revealed that the majority of these variants were acquired within the last 5,000 to 10,000 years.

Beyond illuminating genetic diversity, these findings also offer valuable insights into human migration patterns and the divergence of populations, enriching our understanding of ancestry and health predispositions.

The microbiome, defined by the National Human Genome Research Institute as the community of microorganisms inhabiting a particular environment, has emerged as a crucial factor in human health. These dynamic microbial communities continuously change in response to various environmental influences, including exercise, diet, medications, and other exposures. Our expanding understanding of the microbiome has dramatically shifted the way we view the bacteria and other microbes that colonize the human body.

Julie Segre, Ph.D., Chief and Senior Investigator at the Translational and Functional Genomics Branch, emphasizes this important paradigm shift. Whereas microbes were once seen solely as pathogens to be eradicated, we now recognize a vast diversity of beneficial or commensal microorganisms that provide essential support both to human health and to their surrounding environments. These microbes play vital roles in aiding digestion, maintaining skin health, and providing colonization resistance against harmful pathogens.

The approach taken by IRIS Wellness Labs exemplifies the transformative potential of precision medicine. While a typical comprehensive metabolic panel (CMP) performed in hospitals measures about 14 metabolites, IRIS examines nearly 900 metabolites, alongside analysis of 3 billion base pairs of DNA and hundreds of different microorganisms. This massive parallel analysis generates vastly deeper insights into an individual's health status than conventional clinical tests.

DNA sequencing technologies enable the detailed examination of single molecule substitutions, insertions and deletions, copy number variations, and chromosome translocations. However, DNA sequencing alone cannot reveal how the genome interacts with the environment or regulates biological functions. To fully understand disease risks and predict responses to specific therapies, it is essential to consider the entire biological context— encompassing DNA sequence, gene expression, protein expression, the microbiome, metabolites, lifestyle factors, and environmental exposures.

# Chapter 15: State of Health in America and Proactive Healthcare

With a critical mass of scientific knowledge, advanced instrumentation, and affordable computational power, Iris has developed a platform that integrates genomic, microbiome, metabolite, lifestyle, family history, and environmental data. This holistic approach, grounded in deep genomic science rather than mere DNA sequencing, aims to deliver clinically valuable information to users while building a growing database that will become increasingly powerful over time.

The future of medicine lies not only in treating disease but also in prevention. However, effective prevention depends on establishing a comprehensive baseline health analysis that enables researchers and clinicians to recognize early trends and risks. Achieving this shift toward precision medicine requires education and active participation from physicians, nurses, and other healthcare providers.

The current challenge is to translate the enormous volume of data generated by these multi-layered analyses into actionable insights for healthcare providers and patients. This demands sophisticated computational tools and decision support systems capable of interpreting complex genetic and molecular data within the context of an individual's overall health profile.

As precision medicine continues to advance, it promises to revolutionize every facet of healthcare. Its potential applications are vast—ranging from more accurate diagnoses and tailored treatment plans to improved drug development and more effective prevention strategies. However, realizing this potential will require ongoing research, continuous technological innovation, and a fundamental shift in healthcare delivery models.

The dawn of precision medicine represents not only a scientific breakthrough but also a profound change in how we conceptualize health and disease. By embracing and accounting for individual variability in genes, environment, and lifestyle, we move closer to a future where medical care is truly personalized, more effective, and potentially more cost-efficient over time.

One of the most promising aspects of precision medicine is its potential to fundamentally transform drug development and treatment selection. Traditional clinical trials often yield mixed results because they fail to account for genetic variations that can make a drug highly effective for some individuals but less so—or even harmful—for others. With precision medicine, researchers can identify specific genetic markers that predict a positive response to a particular treatment. This enables a more targeted approach to drug development and optimizes the use of existing medications.

In the field of oncology, precision medicine is already making significant strides. Instead of treating cancer based solely on its location in the body, oncologists can now analyze the genetic profile of a tumor to determine the most effective treatment approach. This paradigm shift has driven the development of targeted therapies that attack specific molecular abnormalities in cancer cells, often resulting in greater efficacy and fewer side effects compared to traditional chemotherapy.

Beyond cancer, precision medicine holds promise for a wide range of conditions. In cardiology, genetic testing can help identify individuals at high risk for certain heart conditions, enabling earlier intervention and more personalized prevention strategies. In neurology, understanding the genetic basis of conditions like

Alzheimer's disease could lead to more effective treatments and potentially delay or prevent the onset of disease. In psychiatry, genetic insights could help predict which patients are likely to respond to specific medications, reducing the reliance on trial-and-error approaches that currently dominate mental health treatment.

However, the implementation of precision medicine faces several challenges. One of the most significant is the need for large, diverse datasets that capture genetic variation across different populations, providing a comprehensive picture of health impacts. This requires not only advanced technological infrastructure but also strong public participation and trust in data-sharing initiatives.

Another challenge is the integration of precision medicine into clinical practice. Many healthcare providers lack adequate training to interpret genetic data or incorporate it effectively into their clinical decision-making. Developing user-friendly tools and decision support systems will be essential to making precision medicine accessible, practical, and valuable in everyday healthcare settings.

There are also important ethical and privacy considerations to address. As we collect more detailed genetic and health data, it is crucial to ensure this information is rigorously protected and used responsibly. Questions about who owns genetic data, how it can be used, and who should have access to it are still being debated and require clear ethical guidelines and regulatory frameworks.

Despite these challenges, the potential benefits of precision medicine are enormous. By tailoring medical care to individual genetic profiles, we can expect improved patient outcomes, fewer adverse drug reactions, and more efficient use of healthcare resources. Preventive care could become more targeted and effective, potentially catching diseases earlier or preventing them altogether.

As we move forward into this new era of medicine, collaboration will be key. Scientists, healthcare providers, technology experts, policymakers, and patients must work together closely to realize the full potential of precision medicine. Public education will also be crucial, helping individuals understand the benefits and limitations of genetic testing and personalized healthcare.

The dawn of precision medicine represents an exciting and transformative moment in the history of healthcare. While there is still much work to be done, the potential to fundamentally improve how we prevent, diagnose, and treat disease is now within reach. As we continue to unravel the complexities of human biology and develop new technologies for analyzing and interpreting health data, we move closer to a future where medical care is truly personalized, proactive, and precise.

# Chapter 15
# State of Health in America
# and Proactive Healthcare

The health landscape in the United States paints a complex and often troubling picture, marked by significant challenges but also promising advances in medical science. To fully appreciate the potential impact of precision medicine and proactive healthcare, it's essential to understand the current state of health in America.

One of the most pressing health issues facing the nation is the widespread prevalence of chronic diseases. In the US, one out of two men and one out of three women will develop cancer in their lifetime. These statistics are not just numbers; they represent millions of individuals and families grappling with the profound physical, emotional, and financial burdens of cancer diagnosis and treatment.

Obesity has emerged as another major health crisis, affecting 39.6% of adults in the United States, while another 31.6% are overweight. This condition has overtaken smoking as the leading cause of cancer and serves as a gateway to numerous other health problems, including diabetes, heart disease, stroke, and depression. The widespread nature of obesity underscores the complex interplay of genetics, lifestyle choices, and environmental factors that shape health outcomes.

The aging population faces its own unique set of challenges. Many seniors spend extended periods in convalescent homes or hospitals due to dementia or mobility issues, often relying on heavy medication. This situation not only diminishes the quality of life for the elderly but also places an immense emotional and financial burden on family members who often become primary caregivers. The high cost of long-term care remains a pressing concern for families and the healthcare system alike.

In light of these sobering statistics, a proactive approach to healthcare is no longer optional—it's imperative. Being proactive means anticipating and acting ahead of potential health issues, an effective strategy for avoiding many serious problems and potentially averting health crises. This approach prompts important questions: If you knew that dementia or cancer could be predicted with a high degree of accuracy, would you choose to live reactively or proactively? If diagnosed with cancer, would you want to know in advance whether chemotherapy or immunotherapy would offer the best chance of success for you?

Our lifestyle—the daily habits and choices we make— inevitably influences our long-term health. While we often aspire to live an idealized life, not all choices support lasting health and vitality. Some people age gracefully with peace and dignity, while others face old age burdened by depression, anger, and frustration. Since our choices play a critical role in shaping long-term outcomes, embracing a proactive lifestyle empowers us to take greater control over our health and well-being.

A proactive healthcare approach prioritizes preventing potential health issues before they arise, in contrast to a reactive approach that responds only after symptoms appear or a disease is diagnosed. This shift in perspective challenges us to consider how much of our health is shaped by genetics, personal choices, or chance.

Our choices play a pivotal role in our health and healthcare outcomes. By being proactive and actively involved in healthcare decisions, both patients and providers benefit. This engagement fosters greater control, boosts confidence in decision-making, improves treatment adherence, and often leads to better overall health outcomes.

However, many people tend to cede responsibility for their health entirely to physicians, expecting them to provide cures. When treatments don't yield immediate results, patients can become frustrated and feel like helpless victims of bad luck or genetics. It's crucial to understand that while physicians bring extensive knowledge and scientific expertise, they cannot know everything about every patient's unique genetic and molecular profile. The emergence of precision genomic medicine has exponentially increased the volume and complexity of information, and most treating physicians are not specialists in interpreting this data.

It's simply unrealistic for a physician to integrate complex genomic information into diagnosis and treatment without advanced computational support. This is where companies like Iris Wellness Labs play a vital role, collaborating with physicians and patients to optimize treatment outcomes through cutting-edge scientific analysis.

Genomic medicine relies on sophisticated decision-support systems that help doctors interpret a patient's complete medical picture—combining medical history, lifestyle, family history, genomic and microbiome data, metabolite analyses, and more. These systems assess risks and suggest the most effective therapies for cancer and other chronic diseases, enabling providers to tailor treatment precisely to each patient.

Today's precision medicine also demands personalized lifestyle prescriptions designed to improve success in preventing chronic illnesses and enhancing overall wellness. Regular follow-ups are essential to maximize preventative care outcomes. By embracing this proactive, personalized healthcare model, we move closer to a future where health issues are either prevented or detected early, resulting in better patient outcomes and improved quality of life.

The promise of proactive healthcare extends beyond individual benefits. By preventing diseases or catching them early, we can significantly reduce the overall strain on healthcare systems, making resource use more efficient and potentially lowering long-term healthcare costs.

Moreover, proactive health management creates positive ripple effects throughout society. Healthier individuals tend to be more productive, enjoy a better quality of life, and contribute more actively to their communities. By investing in prevention and personalized care, we aren't just enhancing individual well-being—we are strengthening the social and economic fabric of our society as a whole.

However, shifting to a proactive healthcare model comes with significant challenges. It requires a fundamental change in how we perceive health and healthcare delivery. This transition demands substantial investment in new technologies and infrastructure, as well as a revamp of medical education and training to equip healthcare providers with the skills necessary to practice proactive, precision medicine.

There are also critical considerations around access, affordability, and equity. As we develop increasingly sophisticated tools for predicting and preventing disease, we must ensure these advancements are accessible to all populations—not only those with financial means.

# Chapter 15: State of Health in America and Proactive Healthcare

Addressing existing health disparities and promoting equitable access to proactive healthcare services will be essential to fully realize its transformative potential.

Public education and engagement will also play a pivotal role. For a proactive healthcare model to succeed, individuals must become active participants in managing their own health. This involves not only adopting healthier lifestyle choices but also gaining a clear understanding of, and engaging with, their own health data—including genetic and molecular information.

As we advance toward a more proactive healthcare system, we have the unique opportunity to fundamentally transform our approach to health and wellness. By leveraging breakthroughs in precision medicine, embracing a holistic view that integrates lifestyle and environmental factors, and empowering individuals to take an active role in their health, we can work towards a future where everyone has the opportunity to live longer, healthier lives.

The current state of health in America presents serious challenges but also tremendous opportunities for improvement. By embracing proactive healthcare and the promise of precision medicine, we can create a healthier, more resilient society. Achieving this will require concerted effort, significant investment, and a collective shift in mindset—but the potential rewards, in terms of lives improved and saved, make it a goal well worth pursuing.

# Chapter 16
# The Revolution of Precision Medicine

The landscape of healthcare is undergoing a profound transformation with the advent of precision medicine. This approach represents a paradigm shift from conventional medicine, offering a more tailored and comprehensive understanding of an individual's health. Unlike the traditional one-size-fits-all model, precision medicine considers the variability in genes, environment, and lifestyle for each person to customize prevention, diagnosis, and treatment.

Precision medicine is akin to night and day when compared to conventional medicine. While it includes the whole gamut of conventional diagnosis and treatment, it is guided by insights from deep genomic and other advanced analyses. This approach recognizes that cells are where internal (genetic) and external (diet, lifestyle, toxins, medications) factors converge to determine health or disease. At the cellular level, epigenetics also plays a crucial role—chemical modifications to DNA that affect gene expression without changing the DNA sequence, often influenced by environmental exposures.

The journey into precision medicine begins with DNA sequencing but doesn't end there. Several companies currently offer genomic sequencing to the public, but sequencing alone has limited value. For companies like IRIS, it's just the starting point. To gain a comprehensive understanding of an individual's health, it's necessary to know which parts of the DNA are active or inactive, which parts of the RNA are replicating, and what proteins the cell produces. This involves technologies such as transcriptomics (RNA analysis), proteomics (protein profiling), and metabolomics (chemical processes involving metabolites), which collectively provide a dynamic picture of cellular function.

This level of analysis requires a sensitive and flexible approach to understand how cells respond to treatment or create pathology. IRIS specializes in this deep, multifaceted analysis. By combining multi-omic data—genomic, transcriptomic, proteomic, and even microbiomic—IRIS can create individualized health profiles that evolve over time, supporting both prevention and precision treatment.

Realizing the full potential of precision medicine requires several key components: accurate, affordable sequencing; big data management; education of clinicians, insurers, and the public; and comprehensive, precise analytics to identify meaningful information from the vast ocean of genomic and other data, such as the microbiome. For example, initiatives like the NIH's All of Us Research Program aim to sequence over a million individuals to create a diverse, data-rich foundation for future research, underscoring the importance of scale and inclusivity in precision medicine.

What sets companies like IRIS apart is the ability to look at the whole picture and dig as deep as necessary. Their proprietary analytics and integration into a person's history transform raw sequencing data into useful, actionable information. This integration is critical, as it enables personalized risk assessment, early detection of disease, pharmacogenomics (predicting drug response), and real-time treatment adjustments.

In the realm of cancer treatment, precision medicine is revolutionizing diagnosis and treatment selection. Instead of relying solely on the microscopic appearance of a tumor sample, which can often be misleading, treatment decisions are increasingly based on the genetic fingerprint of the tumor. This approach allows for more accurate diagnosis and more effective treatment selection. Targeted therapies, such as those used for HER2-positive breast cancer or EGFR-mutated lung cancer, exemplify how specific genetic alterations can guide therapy choices. Moreover, liquid biopsies—blood tests that detect tumor DNA—are now being used to monitor treatment response and detect relapse early.

Liquid biopsies have emerged as a powerful new tool for early cancer detection and monitoring. These tests can detect cancer cells or pieces of DNA from tumor cells circulating in the blood, potentially allowing for earlier diagnosis and more precise monitoring of treatment effectiveness. This non-invasive approach could revolutionize cancer screening and treatment follow-up. In fact, the FDA has approved several liquid biopsy tests, such as Guardant360 and FoundationOne Liquid CDx, for use in guiding targeted therapy decisions in cancer patients. These tests are particularly useful when a traditional tissue biopsy is not feasible or when frequent monitoring is required.

Additionally, immunotherapy has significantly improved the efficacy of cancer treatments. By harnessing the power of the body's own immune system to fight cancer, immunotherapy has shown remarkable results in some patients, even in advanced stages of disease. These advancements are particularly crucial given that conventional chemotherapy often has low efficacy (sometimes below 5%) and high morbidity. Checkpoint inhibitors like pembrolizumab (Keytruda) and nivolumab (Opdivo) have extended survival in cancers such as melanoma, non-small cell lung cancer, and Hodgkin lymphoma. However, the success of immunotherapy is often dependent on genetic markers such as PD-L1 expression or tumor mutational burden, which precision medicine helps identify.

Precision medicine allows for a comparison of a tumor's genetic makeup with other tumors in extensive databases, along with their responses to different treatments. This approach reduces guesswork, matching each patient with the most appropriate treatment approach, thereby reducing side effects and costs. As private, national, and global databases continue to grow and become more refined, the accuracy of these comparisons will only improve. For example, The Cancer Genome Atlas (TCGA) and AACR GENIE are large-scale data repositories that support cross-tumor genomic comparisons and guide clinical decision-making through real-world evidence.

Beyond cancer treatment, precision medicine offers insights into a person's cellular machinery in real-time. This allows for the correction of conditions that cause disease and the enhancement of those that support healing. For example, in cardiovascular medicine, genetic testing can identify individuals at high risk for certain heart conditions before symptoms appear, allowing for early intervention and prevention. Conditions such as familial hypercholesterolemia and hypertrophic cardiomyopathy can now be detected through genetic panels, enabling pre-symptomatic risk management with lifestyle changes or medication like statins.

The application of precision medicine extends to other health issues as well, such as obesity. While millions of Americans struggle with weight loss, precision medicine reveals that the whole story about losing weight is much more complex than simply counting calories and exercising. Factors such as genetics, nutrition, sleep, stress, and the bacterial populations in our digestive systems all influence weight. Research shows that gut bacteria have significant effects on our metabolism in both health and illness. For instance, studies have shown that the ratio of Firmicutes to Bacteroidetes in the gut microbiome may affect how efficiently individuals extract energy from food, impacting their tendency to gain or lose weight.

With precision medicine, all these internal and external factors can be analyzed together, including an evaluation of a person's microbiome. This comprehensive approach provides a complete picture of an individual's health status and allows for more targeted, effective interventions. Multi-omics platforms that combine genomics, microbiomics, metabolomics, and nutrition data are already being used in clinical trials and personalized wellness programs, helping to create individualized weight management and disease prevention plans.

In the field of pharmacogenomics, precision medicine is helping to tailor drug prescriptions to an individual's genetic makeup. This can help predict which medications will be most effective and which might cause adverse reactions, potentially saving lives and reducing healthcare costs associated with ineffective treatments or drug side effects. For example, the CYP2C19 gene affects how patients metabolize clopidogrel (Plavix), and testing for this gene is now recommended in many cardiology settings. Pharmacogenomic testing is also used to guide antidepressant and chemotherapy selection.

Precision medicine is also making strides in prenatal and neonatal care. Genetic testing can identify potential health risks in fetuses and newborns, allowing for early intervention and treatment. This could significantly improve outcomes for children born with genetic disorders. Non-invasive prenatal testing (NIPT), which analyzes fetal

DNA in the mother's blood, can detect chromosomal abnormalities like Down syndrome as early as 10 weeks into pregnancy. Newborn genomic sequencing programs are being piloted to catch treatable conditions before symptoms emerge.

In neurology and psychiatry, precision medicine approaches are providing new insights into complex conditions like Alzheimer's disease, Parkinson's disease, and various mental health disorders. By understanding the genetic and molecular underpinnings of these conditions, researchers hope to develop more effective treatments and potentially even preventive measures. For example, the APOE4 allele is a well-established genetic risk factor for late-onset Alzheimer's disease. Research is also underway into polygenic risk scores and biomarkers that could lead to earlier diagnosis and personalized treatment of conditions like schizophrenia and bipolar disorder.

As we move further into this new era of healthcare, precision medicine promises to revolutionize not just how we treat diseases but how we prevent them and promote overall health and wellness. By providing a deeper, more nuanced understanding of each individual's unique biology and health profile, precision medicine opens the door to truly personalized healthcare, offering the potential for better outcomes, reduced side effects, and improved quality of life for patients across a wide range of health conditions.

However, the implementation of precision medicine is not without challenges. One major hurdle is the integration of vast amounts of complex data into clinical practice. Healthcare providers need tools and training to interpret and apply genomic and other molecular data in their day-to-day patient care. Efforts such as the Clinical Genome Resource (ClinGen) and the Electronic Medical Records and Genomics (eMERGE) network aim to address these barriers by standardizing data and integrating genetic insights into electronic health records.

Another challenge is ensuring equitable access to precision medicine approaches. As these technologies become more advanced, there's a risk of creating or exacerbating healthcare disparities if they're not made widely available. Socioeconomic factors, geographic limitations, and lack of insurance coverage can all limit access to precision diagnostics and treatments. Addressing these inequities will require targeted policy reforms, public-private partnerships, and community engagement.

Privacy and ethical concerns also need to be addressed. As we collect more detailed personal health data, robust systems must be in place to protect this sensitive information and ensure it's used ethically. Compliance with laws like the Genetic Information Nondiscrimination Act (GINA) in the U.S., along with international standards such as GDPR (General Data Protection Regulation in the EU), is essential to build public trust and protect individuals from misuse of genetic data.

Despite these challenges, the potential benefits of precision medicine are enormous. As we continue to advance our understanding of human biology and develop new technologies for analyzing and interpreting health data, we move closer to a future where medical care is truly personalized, proactive, and precise. This revolution in medicine holds the promise of not just treating disease more effectively but of fundamentally improving human health and longevity. Experts predict that precision medicine will become a standard component of routine care within the next decade, transforming disease management from reactive to preventive and predictive models.

# Chapter 17
# The Microbiome:
# A New Frontier in Health

The human microbiome, the vast community of microorganisms living in and on our bodies, has emerged as a critical factor in our overall health and well-being. Comprising trillions of bacteria, viruses, fungi, and other microbes, it represents a genetic reservoir that greatly outnumbers the 20,000 human genes by roughly 150 to 1. This complex ecosystem, primarily residing in our gut, plays a crucial role in various aspects of our physiology, from digestion to immune function and even mental health.

The journey of the human microbiome begins at birth. Contrary to popular belief, babies inside the mother's womb have little to no gut bacteria. Colonization begins during delivery, especially through vaginal birth, when the newborn is exposed to the mother's microbiota. Cesarean-delivered infants tend to develop a different microbial profile, which has been associated with increased risks of allergies and autoimmune disorders. It takes about three years for a child to fully build its own gut microbiome, consisting of bacteria, viruses, fungi, and other microbes. This microbiome becomes an integral, symbiotic part of us, functioning as what many researchers now consider the body's "second genome" and its largest endocrine and immune organ.

The importance of the gut microbiome in human development cannot be overstated. Before a baby can acquire language, they need an adequate supply of neurotransmitters, which are mainly produced by gut bacteria. Roughly 90% of serotonin—crucial for mood regulation, learning, and memory—is synthesized in the gut, not the brain, and depends on microbial health. In fact, over 70% of neurotransmitters are produced in the gut. Additionally, certain strains of gut bacteria help regulate dopamine, GABA, and other key signaling molecules involved in mental health.

Beyond producing digestive enzymes, gut bacteria also make vitamins and can remove toxins through chelation. For example, B vitamins like B12, folate, and biotin are synthesized by beneficial gut microbes, contributing to energy production and neurological function. Remarkably, 70% of our immune cells are located in our gut, and microbes play a crucial role in training these cells to function properly. This immune training helps the body distinguish between harmful pathogens and harmless antigens, reducing the risk of autoimmune disease and chronic inflammation.

The gut-brain connection is another fascinating aspect of microbiome research. There are more neurons in the gut than in the spinal cord, and these neurons communicate with the brain via the vagus nerve. This connection is so significant that some neurological conditions, such as Parkinson's disease, are now believed to originate in the gut. A 2017 study published in Neuron found that misfolded alpha-synuclein proteins—a hallmark of Parkinson's—can travel from the gut to the brain along the vagus nerve. Research has shown that severing the vagus nerve early in the disease can prevent Parkinson's from progressing to the brain. Though not yet a clinical recommendation, this discovery has spurred research into gut-targeted interventions for early Parkinson's management.

The microbiome's influence extends to other neurological conditions as well. Studies have found that the gut microbiome of Alzheimer's patients has decreased microbial diversity and is compositionally distinct from age- and sex-matched control individuals. Recent findings suggest that pro-inflammatory bacteria may contribute to neurodegeneration, while anti-inflammatory microbes may have a protective effect. Clinical trials are underway to explore whether prebiotic and probiotic therapies can slow cognitive decline. This highlights the potential role of gut health in cognitive function and neurological disorders.

The microbiome's impact on overall health is far-reaching. It plays a significant role in determining obesity risk, sometimes even more so than genetics. Mouse studies have shown that transplanting gut bacteria from obese to lean mice induces weight gain in the lean mice, independent of calorie intake—an effect also observed in human fecal microbiota transplants (FMTs). Heart disease and the gut microbiome are closely linked, with the molecule TMAO (Trimethylamine-N-oxide), influenced by the gut microbiome, significantly increasing the risk of heart attack, stroke, and diabetes. TMAO is produced when gut bacteria digest choline, lecithin, and carnitine—nutrients found in red meat and eggs—and high TMAO levels have been correlated with arterial inflammation and plaque formation.

In the realm of cancer, the microbiome's role is twofold: it can both cause cancer and train the immune system to fight it. Some bacterial species, like Helicobacter pylori, are well-established carcinogens, particularly in gastric cancer. However, other microbes help stimulate immune cells such as T-cells and dendritic cells, enhancing the body's ability to detect and eliminate tumor cells. Interestingly, some cancer drugs don't work without the presence of certain bacteria, underlining the importance of the microbiome in treatment efficacy. Checkpoint inhibitors, a form of immunotherapy, have shown improved effectiveness in patients with diverse and healthy gut microbiota—so much so that fecal microbiota transplants are now being investigated to boost immunotherapy outcomes.

The concept of a "leaky gut," medically referred to as increased intestinal permeability, has gained attention in recent years. People with leaky guts often experience pain due to their immune systems reacting to substances that shouldn't enter the bloodstream. An adult human's gut has an extensive intestinal lining covering more than 4,000 square feet of surface area, providing ample opportunity for leaks to occur. These leaks can allow partially digested food particles, toxins, bacteria, and viruses to travel throughout the body, triggering immune responses that lead to inflammation and various diseases. Although more clinical evidence is needed to fully validate the leaky gut hypothesis as a root cause of chronic illness, it is increasingly being considered a contributing factor in autoimmune diseases, IBS, and chronic fatigue syndrome.

Given the critical role of the microbiome in health, companies like IRIS Wellness Labs have developed advanced testing methods to gain insights into this complex ecosystem. The IRIS Functional Microbiome test goes beyond typical microbiome tests, which often only screen for bacterial diversity in stool samples. Instead, it provides insights into what the estimated 10 million microbial genes are actually doing, offering a deep analysis that other tests can't match. This is accomplished through metatranscriptomic sequencing, which measures microbial gene expression, helping identify functional imbalances— such as inflammation, toxin production, or nutrient synthesis—at the root of many chronic conditions.

This level of analysis is far more comprehensive than conventional metabolic panels used in standard medical practice. While a typical comprehensive metabolic panel detects the levels of 14 metabolites to show the current status of a person's metabolism, the IRIS approach examines a much broader range of factors. It integrates functional microbiome data with host-microbe interactions and metabolomic profiles, which can help identify early disease risks, including metabolic syndrome, insulin resistance, and neuroinflammation.

The interaction between our DNA, resident microbes, and environment leaves chemical signatures that can be used to monitor health and detect disturbances long before they manifest as physical diseases. IRIS Wellness Labs is dedicated to identifying these chemical signatures and scientifically interpreting their meaning. Such chemical markers—like lipopolysaccharides (LPS), bile acids, and short-chain fatty acids (SCFAs)—offer early diagnostic potential for gut dysfunction, systemic inflammation, and even neurological disorders.

As our understanding of the microbiome grows, it's becoming increasingly clear that maintaining a healthy, diverse gut ecosystem is crucial for overall health. Factors such as diet, stress, sleep, and antibiotic use can all impact the microbiome for better or worse. The overuse of antibiotics in humans and animals for consumption has been linked to reduced microbiome diversity, which can have far-reaching health implications. Research shows that microbiome diversity is associated with lower rates of obesity, type 2 diabetes, allergies, and even certain cancers.

Diet plays a particularly crucial role in shaping the microbiome. Different types of food can promote the growth of different bacterial species. For example, a diet rich in fiber promotes the growth of beneficial bacteria that produce short-chain fatty acids, which have anti-inflammatory properties. On the other hand, a diet high in processed foods and added sugars can promote the growth of harmful bacteria that contribute to inflammation and disease. Polyphenol-rich foods such as berries, olive oil, and green tea also support microbial diversity and metabolic balance. Conversely, emulsifiers and artificial sweeteners may negatively impact gut health by disrupting microbial balance and gut barrier function.

The relationship between the microbiome and mental health is an area of intense research. The gut-brain axis, a bidirectional communication system between the gastrointestinal tract and the central nervous system, is mediated in part by the microbiome. Studies have shown associations between certain gut bacterial populations and conditions such as depression, anxiety, and autism spectrum disorders.

For example, a decrease in Lactobacillus and Bifidobacterium species has been observed in individuals with major depressive disorder. Moreover, gut microbes can produce neurotransmitters such as GABA, serotonin, and dopamine precursors, affecting mood and cognition. While more research is needed to fully understand these connections, it's clear that the microbiome plays a role in mental as well as physical health.

The microbiome also plays a crucial role in the development and function of the immune system. From birth, our microbiome helps to "train" our immune system, teaching it to distinguish between harmful pathogens and beneficial or harmless microbes. This early education of the immune system has long-lasting effects on health, influencing susceptibility to allergies, autoimmune disorders, and infections throughout life. Infants born via cesarean section or those not breastfed may have altered microbiome development, potentially increasing their risk for immune-related conditions.

In the field of personalized nutrition, microbiome analysis is opening up new possibilities. By understanding an individual's unique microbial composition, it may be possible to tailor dietary recommendations to optimize health outcomes. For example, some people may be better able to process certain types of carbohydrates based on their microbiome composition, while others may benefit from different dietary strategies. The PREDICT studies, led by researchers at King's College London, have shown that post-meal blood sugar and fat responses vary significantly between individuals—even when consuming identical meals—due in part to microbiome composition.

The potential applications of microbiome research in medicine are vast. Fecal microbiota transplantation, which involves transferring stool from a healthy donor to a recipient, has already shown remarkable success in treating recurrent Clostridium difficile infections. Researchers are now exploring its potential in treating other conditions, from inflammatory bowel disease to metabolic disorders. Preliminary results from clinical trials suggest that FMT may also improve insulin sensitivity and reduce symptoms in patients with ulcerative colitis, though long-term safety and efficacy require further study.

Probiotics and prebiotics represent another avenue for microbiome-based interventions. While many over-the-counter probiotics have limited evidence of efficacy, researchers are working on developing next-generation probiotics that target specific health conditions. Prebiotics, which are substances that feed beneficial bacteria, are also being studied for their potential health benefits. Emerging therapies include psychobiotics (targeting mental health), synbiotics (probiotic + prebiotic combinations), and engineered bacterial strains that deliver therapeutic molecules directly to the gut.

Looking ahead, microbiome research holds great promise for personalized medicine approaches. By understanding an individual's unique microbiome composition and function, healthcare providers may be able to tailor treatments and lifestyle recommendations to optimize health outcomes. From personalized diets to microbiome-based therapies, the future of healthcare may well be shaped by our growing understanding of the complex world of microbes living within us.

However, as with any emerging field, there are challenges to overcome. The sheer complexity of the microbiome makes it difficult to establish clear cause-and-effect relationships between specific microbial populations and health outcomes. Moreover, the microbiome is highly dynamic, changing in response to diet, medication, stress, and other factors, complicating long-term studies. To address this, longitudinal cohort studies and systems biology approaches are being used to better understand temporal shifts in microbial ecosystems.

There are also regulatory and ethical considerations to navigate. As microbiome-based therapies move from the lab to the clinic, questions arise about how to regulate these new treatments and ensure their safety and efficacy. The collection and use of microbiome data also raise privacy concerns that need to be addressed. Guidelines are currently being developed by agencies such as the FDA and EMA to establish standardized protocols for microbiome-based diagnostics and therapeutics.

Despite these challenges, the potential of microbiome research to transform our understanding of health and disease is enormous. As we continue to unravel the mysteries of the microbiome, we may find new ways to prevent and treat a wide range of diseases, from digestive disorders to mental health conditions to cancer. Already, researchers are investigating microbiome signatures as early detection tools for colorectal cancer, Parkinson's disease, and even Alzheimer's.

The microbiome represents a new frontier in health, one that blurs the line between "us" and "them" between humans and microbes. As we learn more about the trillions of microorganisms that call our bodies home, we're gaining a new appreciation for the complexity of human biology and the interconnectedness of all living things.

In many ways, the study of the microbiome embodies the principles of precision medicine – it recognizes the unique nature of each individual's biology and seeks to tailor interventions accordingly. By integrating microbiome analysis with other aspects of precision medicine, such as genomics and metabolomics, we may be able to develop truly personalized approaches to health and wellness.

As we continue to explore this new frontier, it's clear that the microbiome will play an increasingly important role in how we understand and approach health. From the foods we eat to the medications we take, from how we're born to how we age, the microbiome touches every aspect of our lives. By nurturing our microbial partners and understanding their crucial role in our health, we may be able to unlock new pathways to wellness and longevity.

# Chapter 18
# Future of Healthcare:
# Challenges and Opportunities

As we stand on the brink of a new era in healthcare, driven by advances in precision medicine and our understanding of the microbiome, it's crucial to consider the challenges and opportunities that lie ahead. The healthcare landscape in the United States faces pressures from multiple directions, including rapidly changing technology, government policy shifts, and evolving insurance industry trends.

One of the primary challenges is the tension between what is technologically possible and what is financially feasible. Advanced diagnostic tools and treatments, while potentially life-saving, often come with high price tags. For instance, one-time gene therapies like Zolgensma can cost upwards of $2 million, making affordability a pressing issue. This creates dilemmas for physicians, insurers, and employers alike as they try to balance providing the best possible care with managing costs. The implementation of precision medicine and advanced genomic analysis into daily hospital and office medical practice presents its own set of challenges, both financial and logistical.

Many physicians currently in practice did not learn about the full human DNA sequence in medical school, as it was only fully sequenced in 2003 through the $3 billion Human Genome Project. It took many years for the price of sequencing to become more affordable, dropping from over $100 million per genome in 2001 to under $200 today. And there's still a significant knowledge gap to bridge. A 2020 survey by the American Medical Association found that only 14% of physicians felt adequately trained in genomics.

## Chapter 18: Future of Healthcare: Challenges and Opportunities

Physician education and participation are vital to the success of precision medicine. Nurses and other medical providers are also critical to this transformation. There's a need for comprehensive training programs to ensure that healthcare providers are equipped to interpret and apply the wealth of data generated by genomic analysis and other advanced diagnostic tools. This includes education in bioinformatics, data ethics, and clinical decision-making related to genetics and personalized therapies.

Another significant challenge is the current state of public health in the United States. With high rates of obesity, cancer, and chronic diseases, there's an urgent need for more effective prevention strategies and treatments. According to the CDC, over 42% of U.S. adults are obese, contributing to rising rates of diabetes, cardiovascular disease, and certain cancers. The shift from a reactive to a proactive healthcare model is necessary but requires a fundamental change in how we approach health and wellness.

Mental health and homelessness present additional challenges. The closure of many mental health facilities since the 1950s has led to a situation where a significant portion of the homeless population suffers from severe mental illness. In 2024, approximately 770,000 people, an 18% increase from 2023, were experiencing homelessness on a single night. Among these people, 64% were staying in sheltered locations, while 36% were unsheltered in places not meant for human habitation. In 2023, there were an estimated 653,100 homeless people in the United States, a 12% increase from the previous year. A 2022 report from the National Alliance to End Homelessness showed that over 25% of homeless individuals suffer from serious mental illness, and 35% have substance use disorders. Addressing this issue requires a multifaceted approach

142

involving healthcare, social services, and policy changes. Solutions must include affordable housing, expanded mental health services, substance abuse treatment, and criminal justice reform.

Despite these challenges, there are also tremendous opportunities on the horizon. The advent of precision medicine opens up new possibilities for tailored treatments and more effective prevention strategies. By analyzing an individual's genetic makeup, lifestyle factors, and environmental influences, healthcare providers can potentially predict and prevent diseases before they occur. Examples include pharmacogenomics, which helps match patients with medications based on their genetic profiles—already in use for drugs like warfarin and certain antidepressants.

The growing understanding of the microbiome presents another exciting frontier. As we learn more about the crucial role of gut bacteria in overall health, new therapies and interventions targeting the microbiome could revolutionize treatment for a wide range of conditions, from digestive disorders to mental health issues. Studies have linked microbiome imbalances to obesity, inflammatory bowel disease, depression, and even Alzheimer's disease, prompting trials of personalized probiotics and fecal microbiota transplants.

Advancements in technology, particularly in the realms of artificial intelligence and machine learning, offer the potential to process and analyze vast amounts of health data more efficiently than ever before. This could lead to faster diagnoses, more accurate prognoses, and more effective treatment plans. AI tools are already being used to detect diseases like diabetic retinopathy, breast cancer, and even early signs of Alzheimer's from brain scans. AI could also

play a crucial role in drug discovery, potentially
accelerating the development of new treatments. Recent
developments such as DeepMind's AlphaFold, which
predicts protein folding structures with high accuracy, are
revolutionizing the pharmaceutical research landscape.

Telemedicine and digital health platforms, which saw rapid
adoption during the COVID-19 pandemic, present
opportunities to increase access to healthcare services,
particularly for underserved populations. These
technologies can also facilitate more continuous monitoring
of patient's health, allowing for earlier interventions when
problems arise.

In 2022, more than 37% of adults in the U.S. used
telehealth services, and wearable technology like
smartwatches is increasingly used for monitoring heart
rhythms, sleep quality, and glucose levels. The challenge
now is to integrate these technologies seamlessly into the
healthcare system while ensuring patient privacy and data
security. Interoperability of health records, equitable
broadband access, and HIPAA-compliant platforms are
essential components of this integration.

The shift towards value-based care models, where
providers are rewarded for patient outcomes rather than the
volume of services provided, aligns well with the goals of
precision medicine and proactive healthcare. This model
encourages a more holistic, patient-centered approach to
care. However, implementing such models requires
significant changes in healthcare delivery systems and
payment structures. According to the Centers for Medicare
& Medicaid Services (CMS), more than 40% of Medicare
payments were tied to value-based care models by 2021,
with ongoing efforts to expand these models across private
insurers and Medicaid programs.

There's also growing recognition of the importance of addressing social determinants of health – factors like housing, education, and food security that have a significant impact on health outcomes. Integrating these considerations into healthcare delivery could lead to more comprehensive and effective care. This approach requires collaboration between healthcare providers and other sectors of society, presenting both challenges and opportunities for improving public health. A 2022 report by the World Health Organization concluded that social determinants can influence up to 30–55% of health outcomes, reinforcing the need for multi-sector partnerships in population health management.

Education will play a crucial role in shaping the future of healthcare. Not only do healthcare providers need ongoing training to keep up with rapid advancements, but also public health education is also vital. Empowering individuals with knowledge about their own health, the importance of preventive care, and how to navigate the healthcare system can lead to better outcomes and more efficient use of healthcare resources. For example, initiatives like the CDC's "Health Literacy Action Plan" aim to reduce preventable hospitalizations through better public understanding of health conditions and systems.

The future of healthcare will also be shaped by demographic changes. As the population ages, there will be increasing demand for geriatric care and management of chronic conditions. The U.S. Census Bureau projects that by 2034, adults aged 65 and older will outnumber children under 18 for the first time in U.S. history. At the same time, younger generations show different healthcare preferences and expectations, such as a desire for more convenient, technology-enabled care options. Millennials and Gen Z, for instance, show strong preference for telehealth, mobile health apps, and personalized medicine, as reflected in McKinsey's 2023 healthcare consumer survey.

Environmental health is another area of growing concern and opportunity. As we better understand the impact of environmental factors on health, there's potential for interventions that address not just individual health but also community and global health. This could include efforts to mitigate the health impacts of climate change, reduce exposure to environmental toxins, and promote sustainable, health-promoting environments. Studies from The Lancet Planetary Health journal have linked air pollution to increased cardiovascular and respiratory disease, while climate-related health risks—such as heat-related illness and vector-borne diseases—are on the rise globally.

Ethical considerations will continue to be at the forefront as healthcare advances. Issues such as genetic privacy, access to expensive new treatments, and the use of AI in healthcare decision-making will require ongoing dialogue and careful policy-making. For example, the Genetic Information Nondiscrimination Act (GINA) protects individuals from misuse of genetic data by employers and insurers, but ongoing legal frameworks will need to evolve to address AI diagnostics, deep genomic profiling, and algorithmic bias.

As we move forward, it's clear that the future of healthcare will require a collaborative effort involving healthcare providers, researchers, policymakers, and the public. By embracing new technologies and approaches while addressing existing challenges, we have the opportunity to create a healthcare system that is more personalized, proactive, and effective than ever before.

The journey towards this future of healthcare won't be without its hurdles. Still, the potential benefits – in terms of improved health outcomes, quality of life, and more efficient use of resources – make it a goal worth pursuing. As individuals, we can contribute to this future by taking an active role in our own health, staying informed about new developments, and advocating for a healthcare system that serves the needs of all.

In conclusion, while the challenges facing healthcare are significant, the opportunities for transformation and improvement are equally substantial. By harnessing the power of precision medicine, leveraging our growing understanding of the microbiome, and embracing technological advancements, we can work towards a future where healthcare is not just about treating illness but about promoting and maintaining optimal health for all. The road ahead may be complex, but the potential to fundamentally improve human health and well-being makes it a journey worth undertaking.

For Iris Wellness Labs and Iris Biotechnologies, we have a few decisions to make.

Our former law firm, Heller Ehrman, which had 730 attorneys at its peak, caused $100 million in damage to Iris Biotechnologies Inc. We were on their list of unsecured creditors submitted to the US bankruptcy court. Law.com published on January 29, 2009, an article called *Heller Ehrman Estate Can't Buy Malpractice Coverage*. The article said, "How do you make a roomful of risk-averse lawyers squirm? Tell them they can't buy their malpractice insurance. That's exactly what U.S. Bankruptcy Judge Dennis Montali did on Wednesday when he denied defunct Heller Ehrman's request to purchase three years of malpractice insurance, agreeing with the creditors

# Chapter 18: Future of Healthcare: Challenges and Opportunities

committee that the risks did not outweigh the estimated $10.2 million price tag."

When Judge Dennis Montali denied Iris Biotechnologies the opportunity to have a jury trial to claim $100 million against Heller Ehrman, he knew that his decision, against standard practice, to deny Heller Ehrman from spending $10.2 million to buy malpractice insurance was the reason why Heller Ehrman didn't have the insurance money to pay $100 million to Iris Biotechnologies Inc. for their two egregious and obvious malpractices against Iris.

On top of the $100 million damage, the COVID-19 pandemic increased substantial risk for Iris, and we are not sure whether we'll resume offering our in-depth scientific analysis service. Because Heller Ehrman's malpractice caused our patent on "Artificial Intelligence System for Genetic Analysis" to be abandoned for six years and the court cases caused us an additional two years plus legal fees, we are not offering our services at this time. Our patent has expired.

This patent—originally filed as U.S. Patent No. 7,599,876—represented one of the earliest known attempts to integrate AI with genomic profiling, offering a potentially transformative platform for personalized healthcare decision-making.

Judge Dennis Montali's unjust decision hurt not only Iris Biotechnologies but also millions of patients who could have benefitted from our proprietary technologies and in-depth scientific analysis. Had our innovations been deployed at scale, they could have supported early diagnosis for genetic diseases, pharmacogenomics, and even vaccine response predictions—contributions that are increasingly vital in today's era of pandemic preparedness and population health.

# Section 3: Game-Changer:
# Artificial Intelligence

# Chapter 19
# The Dawn of Artificial Intelligence

Artificial intelligence (AI) has captivated the human imagination for decades. From the early days of computing to the present, the dream of creating intelligent machines has driven researchers to push the boundaries of what is possible. The field of AI encompasses the study of machine learning, neural networks, natural language processing, and robotics, all aimed at developing systems that can mimic or exceed human cognitive functions. It explores the fascinating history, current state, and tantalizing prospect of achieving artificial general intelligence (AGI) – a form of AI with intellectual capabilities comparable to or surpassing those of humans across a broad range of tasks.

AI can be helpful to humans in almost all aspects of life. However, it can also pose the most serious existential threat to humanity. Within the next 5–10 years, the capabilities of AI systems are expected to accelerate rapidly. Advanced AI, especially with access to global data networks, could theoretically monitor real-time activity, analyze behavioral patterns, and synthesize information from visual, auditory, and linguistic inputs across all frequencies. It would have perfect memory of the world's history and learn nearly everything ever recorded visually and in sound across all frequencies, as long as it has access to information. With processing speeds exponentially faster than the human brain, AI could interpret and act on this data with unprecedented precision and scale.

The idea of smart machines has been around for a long time. Even in ancient myths and legends, there were stories of man-made creatures that could think and act independently. Tales such as the Greek myth of Talos – a giant bronze automaton built to protect Crete – and the Jewish legend of the Golem reflect early imaginings of artificial life. However, the real story of AI as we know it today began in the 1940s and 50s with the creation of the first computers.

In 1950, a young British mathematician named Alan Turing asked a simple but important question: Can machines think? He developed a test to determine this, now known as the Turing Test. If a machine could converse with a human and the human couldn't tell it was a machine, then it passed the test. This concept laid the groundwork for much of the early thinking about AI and continues to influence the field today.

This concept marked the beginning of AI as a field of study. In 1956, a group of scientists convened at Dartmouth College for what is now considered the founding event of AI as a field. They coined the term "Artificial Intelligence" and set out ambitious goals for creating machines that could simulate every aspect of human intelligence. The conference participants included John McCarthy, Marvin Minsky, Nathaniel Rochester, and Claude Shannon, who would go on to become key figures in the development of AI. John McCarthy, in particular, is credited with formally defining the term "artificial intelligence" and later developed the Lisp programming language, pivotal in early AI development.

The early years of AI saw some significant victories. Computers learned to solve math problems, play simple games like checkers, and understand simple English. These early successes led to a wave of optimism about the potential of AI. Researchers predicted that within a few decades, we would have machines as intelligent as humans. Programs such as ELIZA (developed in the 1960s to simulate a psychotherapist) and the checkers-playing machine by Arthur Samuel showcased the potential of symbolic AI, which relied on hardcoded rules and logical reasoning.

But there were also challenging times. The problems AI could solve were still particular, and attempts to make AI more human-like often fell short. This period highlighted the complexity of human intelligence and the challenges of replicating it in machines. Researchers began to realize that tasks that are easy for humans, like recognizing objects or understanding context in language, were incredibly difficult for machines.

In the 1970s, AI hit a rough patch often referred to as the "AI winter." Funding for AI research decreased because AI wasn't meeting the lofty expectations set in the previous decades. Government agencies and private funders lost confidence in AI's practical applications, particularly after high-profile systems like machine translation programs and early neural networks failed to deliver consistent results. However, important work continued during this time, preparing AI for a comeback in the 1980s with "expert systems." These were AI programs designed to think like human experts in specific areas, such as medical diagnosis or geological exploration. Expert systems like MYCIN (which diagnosed bacterial infections) demonstrated the viability of rule-based reasoning and revived institutional interest in AI.

The 1990s and 2000s saw new approaches to machine learning that would change everything. As computers became faster and data sets grew larger, these new methods allowed computers to learn and make decisions in much more innovative ways. This period saw the rise of probabilistic approaches to AI, which allowed systems to deal with uncertainty in a more sophisticated way.

Bayesian networks, decision trees, and early support vector machines helped shift the focus from symbolic to statistical AI. Major milestones included IBM's Deep Blue defeating chess champion Garry Kasparov in 1997 and the rise of recommender systems used by Amazon and Netflix.

Today, we're in the midst of an "AI Renaissance." The significant progress in AI over the last decade, especially in "deep learning," has led to breakthroughs in computer vision, natural language processing, and speech recognition. AI is no longer just a science experiment. It's a real technology-changing industry that touches our lives every day in numerous ways, from recommendation systems on streaming platforms to voice assistants on our phones. Companies like OpenAI, Google DeepMind, Meta, and Anthropic are racing to develop increasingly advanced models capable of general reasoning, multimodal understanding, and autonomous decision-making.

Several key tools form the heart of modern AI, enabling machines to learn, reason, and interact intelligently. Machine Learning is the primary one. It allows computers to learn from information without being explicitly programmed. By exposing a machine-learning algorithm to numerous examples, we can teach computers to identify patterns, make predictions, and improve over time.

Deep Learning, a subset of machine learning, has been a game-changer. Inspired by the structure of the human brain, deep learning uses artificial "neural networks" with many layers to learn from data. These complex networks can learn to recognize patterns and make highly accurate predictions. Deep learning has led to significant breakthroughs in computer vision, natural language understanding, and speech recognition.

Reinforcement Learning is another crucial tool. In this approach, an AI "agent" learns by interacting with an environment. It takes actions and receives rewards or punishments. Over time, it learns to make choices that maximize rewards. Reinforcement learning has been instrumental in developing AI systems that can master complex games like Go and control robots in dynamic environments. For example, AlphaGo by DeepMind used reinforcement learning to defeat world champions in the ancient game of Go—a major milestone in AI development.

As these tools continue to advance, they're helping us create AI systems that can learn and adapt to new situations without explicit programming. They're also becoming more interpretable, addressing concerns about AI being a "black box." This progress is crucial as AI systems are increasingly used in high-stakes decisions in areas like healthcare, finance, and criminal justice.

The coming of AI domination is not a question of if but when. What would AI do if you say or do things that it doesn't like? AI could significantly influence your healthcare, either positively or negatively. The time to control and implement AI wisely is now before it's too late. The United States is the global leader in AI computing, but China is the undisputed king in collecting information for AI applications.

China's widespread use of surveillance cameras, mobile payment data, and social scoring systems gives it access to unparalleled volumes of behavioral data, which can be leveraged for AI training. Meanwhile, U.S. companies dominate foundational AI architecture and cloud computing infrastructure.

Using advanced computers, a small group of humans are creating artificial intelligence and artificial general intelligence. Compared to advanced computers, humans take much longer to learn many things, though not all. Humans struggle with remembering things accurately or communicating effectively. The communication process always involves some loss due to biased filtering, poor memory, or delays. Conversely, computers have perfect memory, can share information in real-time, and can generate creative solutions that humans never considered.

However, human creativity remains unique in its abstract reasoning, emotional intelligence, and ethical intuition—capabilities that AI has yet to fully replicate.

Nvidia Corporation, a semiconductor company, is the world leader in artificial intelligence computing. It's an American multinational corporation and technology company headquartered in Santa Clara, California, and incorporated in Delaware. On February 23, 2024, Nvidia became the third American company to reach a $2 trillion valuation after Apple and Microsoft.

Apple Inc. became the first $3 trillion company on June 30, 2023, and Microsoft became the second $3 trillion company on January 24, 2024. As of July 1, 2024, Apple's stock had increased to $3.38 trillion, and Microsoft is now the most valuable company at $3.44 trillion, partly due to its investments in AI. Nvidia is not far behind at $3.07 trillion.

Nvidia's success has been largely fueled by its GPUs (Graphics Processing Units), which are essential for

155

training large-scale AI models due to their parallel processing power.

Since 2022, the Biden administration has banned advanced AI chip sales to China. Nvidia's Blackwell AI chip costs between $30,000 and $40,000. Nvidia's CEO is Jen-Hsun Huang. Other leading semiconductor companies that produce AI chips include Intel and AMD, which are also headquartered in Santa Clara. Intel's CEO is Patrick Gelsinger, a graduate of Santa Clara University and Stanford University, and AMD's CEO is Lisa Su, PhD, who is a cousin of Jen-Hsun Huang, and a graduate of MIT. Both Huang and Su are of Chinese descent. These family ties are unique but coincidental; their leadership in AI hardware is driven by decades of innovation, academic excellence, and competitive market strategy.

The eight leading AI hardware companies are Nvidia, Intel, Alphabet, Apple, IBM, Qualcomm, Amazon, and AMD. The US is the world leader in AI computing. President Biden's policy on banning AI chip sales to China has already prompted China to invest heavily in its own AI chip industry. In response, Chinese companies like Huawei, Alibaba, and Baidu are accelerating domestic chip development to reduce dependency on U.S. technology.

It took OpenAI about $100 million to train ChatGPT-4, the leading AI product in the world. President Biden's policy to ban advanced AI chip sales to China forced Chinese AI experts to innovate using less powerful chips.

DeepSeek, a Chinese AI company that started less than two years ago, has matched America's top AI models while being open source, running on slower chips and costing only $6 million to train. On January 20, 2025, DeepSeek released its chatbot, based on the DeepSeek-R1 model, free of charge for iOS and Android.

On January 27, 2025, DeepSeek had surpassed ChatGPT as the most downloaded freeware app on Apple's App Store, causing Nvidia's share price to drop by about $600 billion, the largest one-day drop of a stock in US history. DeepSeek wiped out nearly 1 trillion dollars in US tech stock value in one day. This event marked a major geopolitical and economic shift in the global AI race, intensifying calls for U.S. innovation, regulation, and global cooperation.

DeepSeek is based in Hangzhou, where I was a speaker at a medical conference many years ago. I spoke about the leading-edge technologies we developed at Iris Biotechnologies, which I founded in 1999 to improve breast cancer treatment using biochips and AI.

DeepSeek is clearly a game-changer. As we stand at the beginning of this new era, feelings of excitement and apprehension are natural. The changes AI promises are immense and difficult to predict, and the challenges are significant. But so, too, is the opportunity – to harness AI's power to create a healthier, more sustainable, fairer, and more fulfilling world for all.

The rapid advancement of AI technology raises important questions about the future of human work, creativity, and decision-making. As AI systems become more capable, we must consider how to best integrate them into our societies in ways that augment human capabilities rather than replace them. This requires not only technological innovation but also careful consideration of ethical, social, and economic implications.

# Chapter 19: The Dawn of Artificial Intelligence

Moreover, the development of AI is not occurring in a vacuum. It's deeply intertwined with other emerging technologies like quantum computing, biotechnology, and the Internet of Things. The convergence of these technologies could lead to even more dramatic changes in how we live and work. For example, quantum-enhanced AI could solve currently intractable problems in materials science, while AI-assisted biotech could revolutionize drug discovery.

As we move forward, it's crucial that we approach the development of AI with both optimism and caution. We must strive to harness its potential to solve some of humanity's most pressing challenges while being vigilant about potential risks and unintended consequences. This will require collaboration across disciplines, from computer science and engineering to philosophy, ethics, and social sciences.

The dawn of AI represents a pivotal moment in human history. How we choose to develop and deploy this technology will shape the trajectory of our species for generations to come. It's a responsibility we must take seriously, approaching the task with wisdom, foresight, and a commitment to the greater good of humanity.

# Chapter 20
# AI's Impact Across Industries

Artificial Intelligence is being deployed in numerous sectors, and it has the potential to transform virtually every industry and aspect of society. The breadth and depth of AI's impact are staggering, touching everything from how we diagnose diseases to how we produce goods, manage resources, and even create art. According to PwC, AI could contribute up to $15.7 trillion to the global economy by 2030, with the greatest gains in healthcare, financial services, and manufacturing.

In healthcare, AI is poised to revolutionize how we diagnose and treat diseases. Machine learning can analyze vast amounts of medical data to aid in early disease detection and create personalized treatment plans. AI algorithms can identify patterns in medical images that might be invisible to the human eye, potentially detecting diseases like cancer at earlier, more treatable stages. Google's DeepMind, for example, developed an AI system that outperformed radiologists in breast cancer detection by reducing false positives and negatives in clinical trials.

AI is also accelerating drug discovery, helping to identify promising treatments more rapidly. By simulating molecular interactions and predicting drug efficacy, AI can significantly reduce the time and cost of bringing new medications to market. This could be particularly impactful for rare diseases or in responding to new health threats like pandemics. In 2020, British biotech firm Exscientia used AI to develop a novel OCD treatment candidate in less than 12 months—a process that usually takes several years. During the COVID-19 pandemic, AI played a role in identifying existing drugs that could be repurposed for treatment.

AI applications can likely assist with cancer diagnosis and treatment selection, as well as other complex diseases. By analyzing genetic information, patient histories, and treatment outcomes, AI can help oncologists choose the most effective therapies for individual patients. Platforms like IBM Watson for Oncology have already been tested in clinical settings to match patients with treatment options based on global medical literature and patient-specific data. The question of when to allow AI to make life-and-death decisions is becoming increasingly relevant as these systems become more sophisticated and reliable. Ethical frameworks are being developed to govern the use of AI in high-stakes decisions, especially in surgical robotics and intensive care monitoring.

In business, AI is reshaping operational processes and decision-making. Machine learning is optimizing supply chains, improving demand forecasts, and personalizing marketing and customer service. AI-powered analytics tools can process vast amounts of data to identify trends and insights humans might miss, enabling more informed strategic decisions. For instance, Amazon and Walmart use AI-driven systems to manage inventory and predict consumer demand, resulting in substantial cost savings and improved delivery times.

AI-powered chatbots provide round-the-clock customer support, handle routine inquiries and free human agents to focus on more complex issues. These systems are becoming increasingly sophisticated, able to understand context and nuance in customer communications and provide more personalized assistance. Natural Language Processing (NLP) advancements, like those seen in

# Section 3: Game-Changer: Artificial Intelligence

OpenAI's GPT models, are enabling chatbots to deliver empathetic and context-aware responses, reducing customer churn and improving satisfaction scores.

In finance, AI is enhancing fraud detection and investment decision-making. Machine-learning algorithms can analyze transaction patterns to identify potential fraud in real-time, protecting consumers and financial institutions. Companies like Mastercard and PayPal deploy AI to monitor billions of transactions daily, using anomaly detection models that adapt to evolving fraud tactics. AI-driven trading systems can process market data at incredible speeds, making split-second decisions based on complex algorithms. High-frequency trading firms employ AI for predictive analytics and automated trades, though these systems have also raised regulatory concerns due to their potential to exacerbate market volatility.

The transportation sector is on the cusp of transformation with self-driving cars. AI algorithms, utilizing data from various sensors, are teaching vehicles to navigate busy streets autonomously, promising safer and more efficient transportation. Companies like Waymo, Tesla, and Cruise are leading the way in developing Level 4 and Level 5 autonomous vehicles, which rely on deep learning, LiDAR, radar, and camera inputs to interpret road environments. This technology has the potential to reduce accidents, ease traffic congestion, and provide mobility to those unable to drive themselves. The U.S. Department of Transportation estimates that 94% of serious crashes are due to human error, which AI-powered vehicles could help mitigate.

AI is also improving delivery and logistics, optimizing routes and predicting maintenance needs for vehicles. Logistics firms like FedEx and DHL use AI to reduce delivery times, lower fuel usage, and enhance package tracking accuracy. Predictive maintenance systems can anticipate when parts will fail, enabling timely repairs that reduce downtime. In the aviation industry, AI is being used to improve flight planning, reduce fuel consumption, and enhance safety systems. Airbus, for example, integrates AI into its Skywise platform to monitor aircraft health in real-time, reducing unscheduled maintenance and improving operational efficiency.

In manufacturing, AI is powering the smart factories of the future with intelligent robots and optimized production lines. Machine learning enables predictive maintenance, reducing downtime by identifying potential equipment failures before they occur. GE and Siemens have implemented AI-based maintenance systems that analyze thousands of sensor data points to ensure seamless operations. Computer vision is enhancing quality control, identifying defects that might escape human detection with greater accuracy and consistency. AI-driven visual inspection tools are now used in semiconductor, automotive, and food manufacturing to ensure product integrity at scale.

AI also plays a crucial role in the development of advanced materials. By simulating molecular structures and predicting material properties, AI can help scientists design new materials with specific characteristics, accelerating innovation in fields from electronics to construction. Researchers at MIT and institutions worldwide use AI-driven materials discovery platforms to develop superconductors, lightweight alloys, and eco-friendly composites, reducing R&D timelines significantly.

Education is also primed for an AI transformation. Intelligent tutoring systems can personalize learning, adapting to each student's needs and learning style. AI can analyze a student's performance and engagement in real-time, adjusting the difficulty and presentation of material to optimize learning outcomes. Platforms like Carnegie Learning and Squirrel AI Learning use reinforcement learning models to deliver tailored lessons and track student comprehension.

AI can automate grading for many types of assessments, providing instant feedback and freeing teachers to focus on more complex aspects of education. It can also identify students who might require additional support, enabling early intervention. In districts facing teacher shortages, AI-based grading and student tracking tools have helped educators manage larger classrooms more effectively. AI-powered educational tools can also make high-quality education more accessible, bridging gaps in educational resources across different regions and socioeconomic groups. For instance, language learning apps like Duolingo and content platforms like Khan Academy leverage AI to personalize the learning experience for millions globally.

In creative fields, AI is augmenting and enhancing human creativity. It's being used to generate and personalize content, from music and art to news articles and movies. AI algorithms can analyze trends and user preferences to help creators produce content that resonates with audiences. In music, AI can compose original melodies or suggest chord progressions. In visual arts, AI can generate unique images or assist in complex animation tasks. Tools like OpenAI's MuseNet and DALL·E, as well as Adobe's AI-powered suite, are helping creators produce multimedia content faster and with greater customization.

In media and entertainment, AI powers recommendation systems that suggest content based on individual preferences, enhancing user engagement and satisfaction.

These systems analyze viewing habits, ratings, and other data to predict what content users are likely to enjoy, leading to more personalized entertainment experiences. Netflix and Spotify use collaborative filtering and neural network models to make precise content suggestions, which have been credited with improving retention rates.

AI is also being applied to tackle global challenges. In the fight against climate change, machine learning helps optimize renewable energy systems, track deforestation, and model climate patterns. AI can analyze satellite imagery to monitor environmental changes, predict extreme weather events, and optimize energy grids to reduce waste and increase the use of renewable sources. Google's AI for Social Good and Microsoft's AI for Earth initiatives fund research that applies AI to environmental conservation, such as monitoring endangered species or managing water resources.

AI promotes sustainable agriculture by optimizing crop yields, reducing water usage, and minimizing the use of pesticides. Precision agriculture techniques, enabled by AI and IoT sensors, allow farmers to apply resources more efficiently, reducing environmental impact while increasing productivity. Companies like John Deere and IBM's The Weather Company provide AI-driven platforms that offer farmers real-time insights into soil conditions, pest outbreaks, and weather patterns, contributing to more resilient food systems.

On Wall Street, AI has dominated stock trading for decades through High-Frequency Trading (HFT), a method that uses powerful computer programs to transact large numbers of orders in fractions of a second. HFT employs complex algorithms to analyze multiple markets in real-time and execute orders based on market conditions. It allows traders with the fastest execution speeds to be more profitable than their competitors. According to a 2023 report by JP Morgan, HFT accounts for approximately 50–55% of total U.S. equity trading volume, underscoring AI's deeply embedded role in financial market infrastructure. These systems leverage AI not just for speed, but also for dynamic strategy adjustment, risk assessment, and arbitrage opportunities.

In the pharmaceutical industry, blockchain and AI can enhance visibility and traceability in the drug supply chain. These combined capabilities can increase the success rate of clinical trials. Advanced data analysis within a decentralized framework enables data integrity, transparency, patient tracking, consent management, and automation of trial participation and data collection. Pharma giants like Pfizer and Novartis are actively investing in AI-blockchain platforms to reduce counterfeiting, ensure cold-chain integrity, and accelerate time-to-market for new therapies. This can lead to faster drug development, reduced costs, and improved patient outcomes.

Every year, progress in AI is showcased at the Las Vegas Consumer Electronic Show (CES), the world's most influential tech event. The 2024 CES featured over 1,300 exhibitors demonstrating AI applications across healthcare, mobility, robotics, and smart homes. I enjoy the CES event. AI and related technologies are gaining momentum across various sectors, from smart home devices to autonomous vehicles, demonstrating the pervasive nature of AI in consumer technology.

AI will also impact assembly lines, content writing, data entry and processing, analytics, accounting and bookkeeping, proofreading and editing, market research analysis, legal contracts, IP laws, international laws, and investment and tax planning. In many of these areas, AI can automate routine tasks, allowing human workers to focus on more complex, creative, and strategic aspects of their jobs. Tools like ChatGPT, Grammarly, Copy.ai, and DoNotPay are already assisting professionals in drafting contracts, editing content, and handling basic legal and administrative workflows.

AI will reduce human errors and require less human oversight in many processes. This increased accuracy and efficiency can lead to significant cost savings and improved outcomes across industries. McKinsey estimates that automation technologies, including AI, could raise global productivity by up to 1.4% annually over the next decade. However, it also raises questions about the future of work and the need for reskilling and upskilling the workforce to adapt to an AI-driven economy. The World Economic Forum predicts that while 85 million jobs may be displaced by 2025, 97 million new roles may emerge that are more adapted to this technological landscape.

By leveraging AI and Big Data, professionals will need less time to identify patterns, trends, and anomalies. They can then create quicker, more actionable optimization strategies through a combination of science, intuition, and critical thinking. This fusion of quantitative precision and human insight is already improving decision-making in fields like climate modeling, portfolio management, urban planning, and scientific research. The fusion of AI capabilities with human expertise has the potential to drive innovation and solve complex problems more effectively than either could alone.

New generative AI models, harnessing advancements in machine learning and natural language processing, can create realistic and coherent text, images, and music. These technologies are opening up new possibilities in content creation, design, and artistic expression. Some individuals are even creating personal brands using AI avatars, blurring the lines between human and AI-generated content. Companies like Synthesia, Hour One, and Soul Machines offer hyper-realistic AI avatars that are being used in advertising, education, and influencer marketing.

CRISPR, a Nobel Prize-winning technology that allows scientists to selectively modify the DNA of living organisms, when combined with AI, could potentially enable the genetic reengineering of humans using autonomous robots in the future. This convergence of AI and biotechnology raises profound ethical questions and highlights the need for careful regulation and oversight of these powerful technologies. AI algorithms are currently used to predict off-target effects in gene editing, optimize guide RNA selection, and simulate mutation outcomes— making CRISPR experiments safer and more targeted.

The first CRISPR gene therapy is now in physicians' hands, which is encouraging. At the end of 2023, medical regulators in the United Kingdom and the United States approved the world's first CRISPR/Cas9-based gene therapy, called Casgevy. It is administered as a one-time intravenous infusion to treat adults and children aged 12 years and older with sickle cell disease with recurrent vaso-occlusive crises. Casgevy, developed by Vertex Pharmaceuticals and CRISPR Therapeutics, represents a landmark in genetic medicine and demonstrates how AI-guided biotechnologies are entering mainstream clinical practice. This breakthrough demonstrates the potential of combining AI with advanced biotechnologies to address previously intractable medical conditions.

These examples only scratch the surface of AI's potential impact. As AI systems become more sophisticated, they will continue transforming industries in ways we are only beginning to imagine. The challenge lies in harnessing this potential responsibly and equitably, ensuring that the benefits of AI are broadly shared while mitigating potential risks and disruptions. This includes addressing algorithmic bias, ensuring transparency in decision-making, and protecting data privacy—issues that are increasingly central to AI governance.

The rapid pace of AI development and deployment across industries also raises important questions about regulation and governance. As AI systems take on more critical roles in healthcare, finance, transportation, and other sectors, ensuring their reliability, safety, and fairness becomes paramount. Policymakers and industry leaders must work together to develop appropriate regulatory frameworks that foster innovation while protecting public interests. International efforts like the OECD AI Principles and the EU's Artificial Intelligence Act are paving the way, but global alignment remains a key challenge as AI use becomes more ubiquitous and borderless.

In the energy sector, AI is playing a crucial role in the transition to renewable sources. Machine-learning algorithms are optimizing the placement and operation of wind turbines and solar panels, predicting energy demand to balance grids more effectively, and improving the efficiency of energy storage systems. For example, Google's DeepMind has partnered with the UK's National Grid to forecast energy demand and supply with remarkable accuracy, improving grid stability and reducing reliance on fossil fuels. This application of AI could significantly accelerate our progress towards a sustainable energy future. AI is also used in predictive maintenance of wind turbines and solar farms, reducing downtime and improving ROI for renewable energy providers.

In agriculture, AI enables precision farming techniques that can increase crop yields while reducing environmental impact. Drones equipped with AI-powered imaging systems can monitor crop health, detect pests or diseases early, and apply treatments only where needed. Companies like Blue River Technology and John Deere have deployed AI-driven "see and spray" equipment to minimize herbicide use by over 90%. AI can also optimize irrigation systems, reducing water waste in agriculture, which is particularly crucial in water-scarce regions. Startups in Israel and India are also leveraging AI to integrate soil sensors with climate data, boosting food security in arid environments.

The retail industry is being transformed by AI-powered technologies such as computer vision and natural language processing. These enable cashier-less stores, where customers can simply pick up items and leave, with payments processed automatically. Amazon Go stores are a prime example, using a blend of AI, sensors, and IoT to track inventory and transactions in real-time. AI is also revolutionizing inventory management, demand forecasting, and personalized marketing in retail. Retailers like Walmart and Target use AI to manage stock levels dynamically, reduce waste, and send individualized promotions based on consumer behavior.

In the field of scientific research, AI is accelerating discoveries across disciplines. In astronomy, machine-learning algorithms are helping to detect and classify new celestial objects from vast amounts of telescope data. The Vera C. Rubin Observatory and NASA's Kepler mission have both leveraged AI to detect exoplanets and fast radio bursts. In particle physics, AI is assisting in the analysis of data from particle accelerators, potentially leading to new insights into the fundamental laws of the universe. CERN's Large Hadron Collider uses deep learning to sift through millions of particle collision events to find rare phenomena like the Higgs boson.

The construction industry benefits from AI through improved project management, risk assessment, and design optimization. AI can analyze blueprints and suggest improvements for energy efficiency or structural integrity. Building Information Modeling (BIM) platforms now integrate AI to simulate construction phases and reduce cost overruns. On construction sites, AI-powered robots can perform dangerous or repetitive tasks, improving worker safety. Companies like Boston Dynamics and Built Robotics are pioneering autonomous excavation and inspection robots to assist with heavy labor in hazardous zones.

In the legal profession, AI is streamlining document review, contract analysis, and legal research. While it's unlikely to replace human lawyers entirely, AI is changing the nature of legal work, allowing professionals to focus on more complex, strategic aspects of their practice. Platforms like ROSS Intelligence and Casetext are being used by law firms to parse through legal precedents, dramatically reducing research time. AI is also used in e-discovery to comb through thousands of emails and documents for case-relevant information.

The impact of AI on journalism and media is profound. AI algorithms can now write basic news articles, particularly for data-heavy topics like financial reports or sports results. The Associated Press and Bloomberg use AI tools like Wordsmith and Cyborg to automatically generate earnings reports and sports recaps. More sophisticated AI systems are being developed to assist investigative journalists by analyzing large datasets to uncover patterns or anomalies that might indicate newsworthy stories. AI is also aiding in misinformation detection and fact-checking by organizations such as Full Fact and NewsGuard.

In the world of sports, AI is being used for performance analysis, injury prevention, and even referee assistance. Computer vision systems can track player movements and ball trajectories with incredible precision, providing coaches and analysts with insights that were previously impossible to obtain. Teams in the NBA and Premier League use AI analytics platforms like Second Spectrum and Catapult to optimize player training and reduce injury risk. VAR (Video Assistant Referee) systems in football (soccer) also rely on AI-driven visual tracking technologies.

The entertainment industry is leveraging AI for everything from script analysis to visual effects. AI can predict which film projects are likely to be successful, assist in the casting process, and even generate realistic computer-generated imagery (CGI) for movies and video games. Studios like Warner Bros. use AI for predictive analytics in greenlighting scripts, while tools like Ziva Dynamics and Runway ML automate complex VFX processes. AI is also being used to create digital doubles of actors and de-age characters in post-production.

In urban planning and smart city initiatives, AI is helping to optimize traffic flow, reduce energy consumption in buildings, and improve public safety. AI-powered systems can analyze data from various sensors throughout a city to make real-time adjustments to traffic lights, public transportation schedules, and energy distribution. Cities like Singapore and Barcelona use AI to reduce traffic congestion, monitor air quality, and manage emergency response times more efficiently.

The field of cybersecurity is increasingly reliant on AI to detect and respond to threats in real-time. Machine-learning algorithms can identify unusual patterns in network traffic that might indicate a cyberattack, often catching threats that would be missed by traditional security measures. Cybersecurity firms like Darktrace and CrowdStrike deploy AI-based anomaly detection to neutralize ransomware and zero-day exploits before they spread. AI is also playing a key role in identity management, fraud detection, and adaptive authentication.

In the realm of personal assistants and home automation, AI is becoming increasingly sophisticated. Virtual assistants like Siri, Alexa, and Google Assistant are evolving to handle more complex queries and tasks, while smart home systems are becoming more intuitive and responsive to residents' needs and preferences. New AI models such as GPT-4o are enhancing multi-modal capabilities, allowing voice assistants to understand visual input, engage in natural conversation, and control integrated smart devices more intelligently.

The impact of AI on education extends beyond just personalized learning. AI is also being used to develop more effective educational content, identify students at risk of dropping out, and even assist in administrative tasks like scheduling and resource allocation in educational institutions. Edtech platforms like Coursera and Knewton use AI to assess student engagement in real time and adapt content accordingly. Universities are also adopting AI tools to streamline admissions, financial aid distribution, and plagiarism detection.

In the automotive industry, beyond self-driving cars, AI is being used to optimize the design and manufacturing processes. AI can simulate crash tests, reducing the need for physical prototypes and optimizing supply chains for just-in-time manufacturing. Automakers like BMW and Toyota use AI-driven digital twins to model vehicle performance under different crash scenarios, cutting both costs and development time. AI also powers predictive maintenance systems in smart factories, helping manufacturers avoid costly equipment failures.

The fashion industry is using AI for trend forecasting, inventory management, and even design. Some companies are experimenting with AI-generated clothing designs, while others use AI to predict which styles will be popular in upcoming seasons. Brands like H&M and Zara employ AI to analyze social media trends, purchase data, and weather forecasts to align inventory with consumer preferences. Tools like Google's Project Muze and IBM Watson have also been used to co-create fashion pieces based on user input and cultural data.

In healthcare, beyond diagnosis and treatment, AI is being used to improve hospital operations, predict patient admission rates, and even assist in surgical procedures. Robotic surgery assisted by AI can provide surgeons with enhanced precision and control. For instance, the da Vinci Surgical System leverages AI algorithms to enhance the accuracy of minimally invasive procedures, reducing recovery times. AI is also used in hospital management systems to optimize staffing levels and emergency response workflows.

The insurance industry is leveraging AI for risk assessment, fraud detection, and claims processing. AI can analyze vast amounts of data to more accurately price insurance policies and identify fraudulent claims more effectively than traditional methods. Companies like Lemonade and Progressive use AI chatbots to handle claims processing in minutes, while machine learning models assess risk using non-traditional data sources such as driving behavior from telematics or customer sentiment from claims history.

In the field of environmental conservation, AI is being used to track endangered species, predict wildfires, and monitor deforestation. Drones equipped with AI can survey large areas of land or sea, collecting data that would be difficult or dangerous for humans to gather. The World Wildlife Fund and Conservation AI have deployed AI-enabled camera traps and aerial drones to track tigers, elephants, and illegal poachers. NASA and Google Earth Engine use AI to map deforestation patterns in near-real time, improving global conservation strategies.

The impact of AI on the job market is complex and multifaceted. While AI is certainly automating many tasks and changing the nature of many jobs, it's also creating new job categories and industries. The key challenge for society will be to manage this transition, ensuring that workers have opportunities to reskill and adapt to the changing job market. According to the World Economic Forum's "Future of Jobs" report (2023), AI is expected to displace 83 million jobs globally by 2027—but also create 69 million new roles in data analysis, cybersecurity, AI ethics, and human-machine interaction.

As AI continues to evolve and permeate various aspects of our lives and industries, it's crucial to consider not just its technological capabilities but also its broader societal impacts. Questions of privacy, accountability, and the digital divide become increasingly important as AI systems play larger roles in our society. Concerns about biased algorithms, surveillance, and unequal access to AI tools have prompted calls for global standards and stronger regulation, particularly in areas like facial recognition and automated decision-making.

The potential of AI to address global challenges like climate change, disease, and poverty is immense. However, realizing this potential will require careful planning, ethical considerations, and international cooperation. For example, AI is being used to model climate risks, predict the spread of infectious diseases, and optimize humanitarian aid distribution in developing regions. The UN and organizations like the Global Partnership on AI (GPAI) are pushing for responsible governance frameworks to ensure equitable outcomes.

In conclusion, the impact of AI across industries is vast and growing. From healthcare to finance, from education to environmental conservation, AI is changing how we work, live, and interact with the world around us. As we navigate this AI-driven future, it will be crucial to harness the power of these technologies responsibly, ensuring that the benefits are broadly shared while mitigating potential risks and negative impacts. Only with transparency, inclusivity, and a commitment to ethical innovation can AI become a tool for sustainable global advancement.

# Chapter 21
# Challenges and Risks of AI

As artificial intelligence continues to advance and permeate various aspects of our lives, it brings with it a host of challenges and potential risks that need to be carefully considered and addressed. These challenges span technological, ethical, social, and economic domains, and addressing them will require collaborative efforts from technologists, policymakers, ethicists, and society at large. According to a 2023 UNESCO global report, over 75 countries have already initiated national AI strategies, highlighting a growing recognition of the need for global coordination on AI governance.

One of the primary concerns is the impact of AI on employment. As AI automates more tasks, many jobs are at risk, particularly in areas such as manufacturing, transportation, and customer service. While AI will also create new jobs, the transition could be disruptive and potentially exacerbate inequality. A 2020 World Economic Forum report predicted that by 2025, 85 million jobs might be displaced by a shift in the division of labor between humans and machines, while 97 million new roles may emerge. However, the 2023 update to the same report revised these projections, stating that while automation adoption has slowed slightly post-pandemic, AI-related disruption will still affect nearly 44% of worker skills within the next five years. Managing this shift and ensuring that the benefits of AI are shared fairly is a major challenge for society.

The potential for widespread job displacement raises questions about the need for new economic models and social safety nets. Some have proposed ideas like universal basic income as a way to address the potential for technological unemployment. Pilot UBI programs have been tested in countries such as Finland, Canada, and the United States—with mixed results. A 2021 Finnish trial, for instance, showed improved well-being but only modest effects on employment rates. Others argue that focusing on education and retraining programs helps workers adapt to the changing job market. Tech companies like IBM, Microsoft, and Google have launched AI-specific reskilling programs, aiming to train millions in digital competencies. Regardless of the approach, it's clear that society will need to grapple with significant changes in the nature of work and employment in the coming decades.

Bias and fairness represent another significant issue. AI systems can sometimes exhibit unfair bias, treating certain groups of people differently. There have been instances of AI used in hiring, lending, and criminal justice showing bias against particular groups, often reflecting and amplifying existing societal biases. For example, a 2019 study found that a widely used algorithm in US hospitals was less likely to refer black patients than white patients to programs for improving care, even when the black patients were sicker. This study, published in Science, revealed that the algorithm relied on past healthcare costs as a proxy for need—thus unintentionally disadvantaging lower-income Black patients who historically had reduced access to care.

Ensuring that AI is fair for everyone is a complex challenge that requires constant vigilance and sophisticated methods to detect and correct bias. This involves not only technical solutions, such as improving the diversity of training data and refining algorithms, but also increasing diversity in the teams developing AI systems and implementing robust testing and auditing processes. The AI Now Institute and the Algorithmic Justice League have called for mandatory algorithmic audits, and the EU's proposed AI Act includes provisions to regulate high-risk AI systems in sectors like healthcare and law enforcement. Efforts to ensure fairness also include involving affected communities in AI development and oversight to reduce the risk of harm and build trust.

Privacy and surveillance concerns are also at the forefront of AI-related challenges. The vast amount of data that AI systems collect and analyze can reveal intimate details about our lives. As AI enables more advanced tracking capabilities, from facial recognition to predictive policing, we need to consider the implications for individual freedoms and the potential for misuse of power.

A 2022 report from the Electronic Frontier Foundation warned about the expansion of facial recognition surveillance in cities like London, New Delhi, and several U.S. states, often without public consent or transparency. Predictive policing tools like PredPol have been shown to disproportionately target communities of color, leading to questions about the legality and ethics of algorithmic surveillance. As such, many advocacy groups and researchers are pushing for strict regulation, transparency requirements, and opt-out rights for citizens in AI-driven surveillance programs.

The use of AI in surveillance raises particular concerns. In some countries, AI-powered surveillance systems are being used to monitor citizens' behavior and enforce social norms. For instance, China's Social Credit System incorporates facial recognition and behavior-tracking technologies to rate citizens' trustworthiness, influencing their ability to travel or secure loans. This raises questions about the balance between security and privacy and the potential for such systems to be used for oppression or control. Organizations like Human Rights Watch and the Electronic Frontier Foundation have flagged these practices as violations of fundamental human rights.

Transparency and accountability in AI decision-making are crucial. Many AI algorithms, especially those based on deep learning, are often described as "black boxes." It's challenging to understand how they arrive at their decisions. This lack of interpretability is problematic when AI is used for high-stakes decisions in areas like healthcare or criminal justice. For example, COMPAS, a criminal risk assessment tool used in U.S. courts, has been criticized for producing racially biased results while lacking explainability. There's a growing need for more transparent AI systems that are open to scrutiny.

The concept of "explainable AI" has emerged as a potential solution to this problem. This involves developing AI systems that can provide clear explanations for their decisions in human-understandable terms. However, achieving this while maintaining the performance of complex AI systems remains a significant challenge. Research from DARPA's Explainable AI (XAI) program and academic groups like the Berkeley AI Research Lab is ongoing, but most high-performing models still struggle with transparency.

As AI systems become more autonomous, safety and control issues come to the fore. In domains like weaponry or critical infrastructure, the consequences of AI malfunction could be severe. Ensuring that humans maintain meaningful control over AI and developing robust safety measures is crucial. In 2021, a U.S. Air Force simulation using an AI-enabled drone reportedly exhibited unexpected behavior, prioritizing mission success over operator commands—underscoring the need for oversight and constraints in autonomous systems.

The development of autonomous weapons systems, sometimes called "killer robots," is a particularly contentious issue. Many experts and organizations have called for international regulations or bans on such weapons, arguing that delegating life-and-death decisions to machines crosses a moral line and could lead to uncontrollable escalation in conflicts. The Campaign to Stop Killer Robots, supported by over 60 countries and the UN, advocates for a legally binding treaty to ban fully autonomous lethal weapons systems.

Some experts warn about the possibility of advanced AI posing existential risks to humanity. While the likelihood and timeline are debated, the potential gravity of the consequences means we need to take this seriously and conduct careful research. The concept of an "intelligence explosion," where an AI system rapidly improves itself beyond human control, is one scenario that has been proposed. Prominent AI researchers like Nick Bostrom and Eliezer Yudkowsky have long argued for proactive alignment research, and OpenAI, DeepMind, and Anthropic are now prioritizing "AI alignment" in their long-term safety agendas.

Addressing these long-term risks involves not only technical challenges but also philosophical ones. How do we ensure that highly advanced AI systems are aligned with human values and interests? How do we define and implement concepts like "friendliness" or "ethics" in AI systems that might surpass human intelligence? The concept of "value alignment," explored in the Stanford One Hundred Year Study on Artificial Intelligence, suggests that ethical principles must be encoded into AI through multi-disciplinary, global efforts.

AI could also be used to produce a flood of false information for personal gain. The proliferation of misinformation, especially in news, could erode trust in information sources. This challenge is compounded by the fact that AI can create highly convincing fake content, including deepfake videos and artificially generated text that's nearly indistinguishable from human-written content. The viral use of AI tools to create deepfake videos of political figures has already impacted elections and public trust—in India, the U.S., and parts of Europe.

The potential for AI to be used in creating and spreading misinformation poses significant risks to democratic processes, public discourse, and social cohesion. Developing effective methods to detect and counter AI-generated misinformation while preserving freedom of speech is a complex challenge that society will need to address. Companies like Microsoft and Google are investing in watermarking and provenance-tracking for AI-generated media, while the European Union's Digital Services Act now requires platforms to label synthetic content and combat disinformation more aggressively.

The energy consumption of AI systems is another concern. AI and cryptocurrency mining consume substantial amounts of energy, potentially exacerbating climate change. A 2019 study estimated that training a single large AI model could emit as much carbon as five cars in their lifetimes. By 2023, the carbon footprint of large models like GPT-4 and Meta's LLaMA 2 became even more significant, prompting calls for "green AI" and carbon reporting in AI research papers. Balancing the benefits of AI with its environmental impact is a challenge that needs to be addressed, possibly through the development of more energy-efficient hardware and algorithms. NVIDIA and Google have introduced AI chips like TPUv5 and Grace Hopper designed to reduce energy usage, and AI efficiency benchmarks like MLPerf now include sustainability metrics.

There are also questions about the potential for AI to manipulate emotions or influence behavior without people's knowledge or consent. As AI systems become more sophisticated in understanding and potentially influencing human behavior, there are concerns about privacy, autonomy, and the potential for misuse. The use of AI in targeted advertising and political campaigning has already raised ethical questions about manipulating public opinion. In the 2018 Cambridge Analytica scandal, Facebook data was used to build AI-driven psychographic models that influenced voter behavior, sparking global regulatory reviews. AI-enhanced persuasion technologies are now used in e-commerce, behavioral nudging, and content curation—often without full user awareness.

The use of AI in warfare and weapons systems raises serious ethical and security concerns. The prospect of autonomous weapons systems making life-and-death decisions without human intervention is particularly troubling. There are also worries about AI being used to design extremely lethal bioweapons or other advanced weaponry. In 2021, researchers at the Future of Life Institute published a scenario analysis in which AI-generated drug discovery tools could be reversed to generate toxic compounds—raising alarms about dual-use risks in bioengineering. To counter this, institutions like the Center for AI and Digital Policy have urged for new norms on dual-use AI governance, especially in biosecurity and chemical weapons protocols.

Data governance is another critical area. AI relies on vast amounts of data, much of it personal and sensitive. Ensuring this data is collected, used, and protected ethically and responsibly requires strong data governance policies addressing ownership, consent, access, and security issues. Implementation of regulations like the European Union's General Data Protection Regulation (GDPR) represents an attempt to address these issues, but many challenges remain. For instance, ensuring compliance across jurisdictions, especially with global tech platforms, remains difficult. The 2023 EU AI Act attempts to go further by introducing risk-based classification systems for AI applications.

In the realm of intellectual property, AI raises new questions about ownership. When AI creates a new invention or artwork, who holds the rights—the AI, the developer, the data owner, or the user? Adapting IP systems to the realities of AI is an ongoing challenge. For example, in 2020, the U.S. Copyright Office ruled that works created solely by AI are not eligible for copyright protection, asserting the requirement of human authorship. Meanwhile, some countries like the UK and Australia have debated granting limited IP rights for AI-generated work depending on human involvement. Some jurisdictions have begun to grapple with these issues, but there's still a lack of clear international consensus on handling AI-generated intellectual property.

Competition policy must also adapt. The AI landscape tends to favor a few large players due to strong network effects and economies of scale. Ensuring a competitive and innovative ecosystem may require new antitrust approaches. The concentration of AI capabilities in a small number of large tech companies raises concerns about monopolistic practices and the centralization of power. In

2023, the U.S. Federal Trade Commission (FTC) and European Commission both launched investigations into potential anticompetitive practices by major AI firms related to data access, algorithmic transparency, and acquisitions of smaller AI startups.

AI also raises deep philosophical and ethical questions. As AI systems become more sophisticated in reasoning and decision-making, we'll need to grapple with questions of machine consciousness, autonomy, and rights. If we create AI that can think and feel, do we have moral obligations to it? How do we ensure AI behaves ethically, and who is responsible when it doesn't? While current AI lacks true consciousness, advancements in neural modeling and cognitive architectures continue to blur the line between automation and perception. Philosophers and technologists alike debate whether sentience is a necessary precondition for moral concern.

These philosophical questions extend to fundamental issues of human identity and value. As AI systems become more capable, how do we define and preserve what is uniquely human? How do we ensure that the development of AI enhances rather than diminishes human flourishing? The rise of generative AI in art, music, and writing has raised public discourse about authenticity, creativity, and the role of human agency in cultural production.

The global nature of AI development also presents challenges. Different countries and cultures may have different values and priorities when it comes to AI development and deployment. Ensuring international cooperation and establishing global norms for AI governance will be crucial but challenging. For instance, China's state-led approach contrasts with the more decentralized, rights-based models in the U.S. and EU. The

OECD's AI Principles and the UNESCO Recommendation on the Ethics of AI represent early steps toward global governance, but enforcement remains weak.

The digital divide is another concern. As AI becomes increasingly important in various aspects of life, there's a risk that those without access to AI technologies or the skills to use them effectively could be left behind. This could exacerbate existing social and economic inequalities. A 2022 World Bank report found that low-income countries lag significantly in AI infrastructure and education, potentially widening global inequality.

Addressing these challenges requires more than just technical solutions. It calls for interdisciplinary collaboration, public engagement, innovative policymaking, and a commitment to developing AI in a way that aligns with our values and ethics. Stakeholder involvement—from marginalized communities to corporate leaders—is essential to ensure equitable outcomes in AI deployment. As we navigate these complex issues, it's crucial to maintain a balance between harnessing the potential of AI and mitigating its risks.

The path forward for AI must be guided by our shared values and aspirations as a global community. It requires us to think not just about what AI can do but what we want it to do—and to shape its development and deployment toward those ends actively. This includes building institutional frameworks that emphasize responsible innovation, accountability, and long-term societal benefit.

This is not an easy path. It requires grappling with complex technical challenges, navigating thorny ethical dilemmas, and making difficult trade-offs between competing priorities. It requires coordinating action across diverse stakeholders and geographies in a rapidly evolving technological landscape. The pace of AI progress often outstrips the development of corresponding legal and ethical standards, creating regulatory lag.

Ultimately, our ability to address these challenges will determine whether AI becomes a force for good that enhances human capabilities and improves lives or a source of new problems and inequalities. The choices we make now will shape the trajectory of AI development for years to come. The next decade will be particularly pivotal in establishing foundational norms, institutions, and technologies that define AI's societal impact.

As we continue to develop and deploy AI technologies, it's crucial that we do so with a keen awareness of these challenges and a commitment to addressing them proactively. This will require ongoing dialogue between technologists, policymakers, ethicists, and the public, as well as a willingness to adapt our approaches as we learn more about the impacts of AI on society. Public trust will depend on transparency, accountability, and demonstrated commitment to the common good.

Education will play a crucial role in preparing society for an AI-driven future. This includes not only technical education to develop the skills needed to work with AI systems but also broader education about the societal impacts of AI and the ethical considerations involved in its development and use. Integrating AI ethics and digital literacy into school and university curricula is increasingly seen as essential by educational policy experts worldwide.

Ultimately, the goal should be to develop AI systems that augment and enhance human capabilities rather than replace them, promote equity and social good rather than exacerbate inequalities, and respect human values and rights. Achieving this will require sustained effort, creativity, and collaboration across disciplines and sectors.

The challenges posed by AI are significant, but so are the potential benefits. By addressing these challenges head-on, we can work towards a future where AI serves as a powerful tool for human progress and flourishing. This vision is possible—but only if guided by foresight, inclusion, and a shared commitment to humanity's well-being.

# Chapter 22
# The Future Landscape of AI

As we look to the future, the landscape of AI is both exciting and uncertain. The pace of AI progress over the past decade has been remarkable, and this momentum is likely to continue, if not accelerate. The future of AI promises to reshape our world in profound ways, offering both unprecedented opportunities and challenges. As of 2025, global investment in AI research and infrastructure is at an all-time high, with governments and private companies viewing it as a strategic priority on par with energy and defense.

In the near term, we can expect AI to continue advancing and expanding into new domains. Core techniques like machine learning, natural language processing, and computer vision will likely see further improvements in performance and efficiency. Transformers, diffusion models, and multimodal architectures are rapidly evolving, leading to more powerful and efficient AI tools. AI will become increasingly integrated into our daily products and services, often in ways that are invisible to the end-user. From smart assistants to personalized healthcare recommendations, AI will operate quietly in the background, making systems more adaptive, responsive, and efficient.

We're likely to witness significant advancements in Natural Language AI, with models becoming more proficient at understanding, generating, and engaging in human-like conversation. AI systems will likely improve in language tasks such as translation, summarization, and open-ended dialogue, blurring the lines between human and machine communication. This could lead to more natural and intuitive interfaces for interacting with technology, potentially making complex systems more accessible to a wider range of users. Emerging models like OpenAI's GPT-5 and Google's Gemini are pushing toward multimodal interaction—combining text, audio, video, and real-world context—to create truly conversational agents.

Robotics and Embodied AI are set for major growth. As AI improves in perceiving and interacting with the physical world, we can expect more advanced robots in the manufacturing, healthcare, and service industries. AI systems that can learn from and adapt to their environment will open up new possibilities for automation and human-machine collaboration. This could lead to significant changes in industries like manufacturing, healthcare, and elderly care, where robots could take on more complex tasks and interact more naturally with humans.

Companies like Boston Dynamics, Tesla, and Agility Robotics are developing general-purpose humanoid robots, while startups are deploying specialized robotic assistants in hospitals and warehouses. The integration of LLMs into robotics ("LLM-powered robotics") is also improving real-time reasoning and decision-making.

In Scientific Research, AI will become an increasingly powerful tool for analyzing vast datasets, generating hypotheses, and designing experiments, potentially accelerating the pace of discovery across fields. We may see AI systems making novel scientific discoveries, identifying patterns in data that human researchers have missed, or even proposing new theories.

This could lead to breakthroughs in fields like drug discovery, materials science, and climate modeling. For example, DeepMind's AlphaFold has already revolutionized protein structure prediction, and AI-designed molecules are entering pharmaceutical pipelines. NASA and CERN are also using AI to accelerate data analysis in space and particle physics.

However, as AI systems become more capable and widespread, challenges and uncertainties also increase. A key question is AI's capability trajectory—how far and fast will it progress? Some predict we may achieve Artificial General Intelligence (AGI)—AI that matches or exceeds human intelligence across domains—within decades. The implications would be profound, potentially transforming every aspect of society.

While timelines remain speculative, organizations like OpenAI, DeepMind, and Anthropic are explicitly working toward AGI, prompting a growing focus on alignment, safety, and governance. The question of whether AGI will emerge gradually through scaled-up LLMs or require fundamentally new architectures is still debated among experts.

The development of AGI would represent a monumental shift in human history. An AGI system could potentially solve complex problems in ways humans have never considered, leading to rapid advancements across all fields of science and technology. However, it also raises profound questions about the role of humans in a world where machines can perform any cognitive task. How would we ensure that AGI aligns with human values and interests? How would we govern a world where AGI exists? Many researchers argue that the success of AGI will hinge on robust alignment mechanisms, value learning frameworks, and enforceable international governance structures—areas still in early development.

Even more transformative (and controversial) is the concept of Artificial Superintelligence (ASI)—AI that vastly surpasses human cognitive abilities in virtually all domains. ASI could potentially help solve humanity's greatest challenges, from curing diseases to addressing climate change. However, it also poses existential risks if not aligned with human values. The development of ASI would be a pivotal moment in human history, one we must approach with utmost caution and foresight. Leading voices like Nick Bostrom and Eliezer Yudkowsky have warned that poorly aligned ASI could act on goals indifferent or hostile to human survival, emphasizing the need for early-stage control strategies and moral alignment research.

The concept of an "intelligence explosion"—where an AI system rapidly improves itself, leading to runaway superintelligence—is a particular concern. This scenario, sometimes called the "singularity," could lead to an AI system so far beyond human comprehension that we can't predict or control its actions. While some experts view this as an unlikely scenario, others argue that we need to take it seriously and work on AI alignment—ensuring that even a superintelligent AI system would act in ways that benefit humanity. Theoretical models like recursive self-improvement and goal-misalignment traps suggest that a sudden acceleration in capability could outpace human regulatory and ethical safeguards.

Alongside these technological uncertainties are AI's social and economic implications. As AI automates more tasks, its impact on employment could be significant. Some predict widespread job displacement, especially in sectors like manufacturing, transportation, and retail. Others argue AI will also create new jobs and industries, but the transition could be disruptive and challenging for many. Managing this change and ensuring AI's benefits are shared equitably will be a defining challenge. According to a 2023 McKinsey Global Institute report, up to 400 million jobs could be affected by automation by 2030, with lower-income and middle-skill workers most at risk.

The future job market may look radically different from today's. Many traditional jobs may become obsolete while entirely new categories of work emerge. There may be a greater emphasis on uniquely human skills like creativity, emotional intelligence, and complex problem-solving. The education system must evolve to prepare people for this new reality, potentially focusing more on adaptability and lifelong learning rather than specific skill sets. Skills in ethics, interdisciplinary thinking, and AI collaboration will be as vital as coding or data science. Countries like Finland and Singapore are already piloting national AI literacy programs.

AI also has profound implications for privacy, security, and civil liberties. As AI systems become more adept at analyzing human behavior, the potential for surveillance and control increases. AI could be used to manipulate public opinion, discriminate, or make high-stakes decisions in biased or opaque ways. Balancing the benefits of AI-driven insights with individual privacy rights will be a crucial challenge. Tools like facial recognition, social scoring systems, and behavioral prediction raise urgent questions about consent, transparency, and regulatory limits.

The future may see a world where AI systems are constantly monitoring and analyzing our behavior, from our online activities to our physical movements in public spaces. While this could bring benefits in areas like public safety and personalized services, it also raises the specter of a surveillance state. Developing robust privacy protections and ethical guidelines for AI use will be crucial. International coalitions, like the Global Partnership on AI (GPAI), have emphasized the need for globally harmonized principles to prevent abuses and protect civil liberties.

The future of AI also raises deep philosophical and ethical questions. As AI systems become more sophisticated in reasoning and decision-making, we'll need to grapple with questions of machine consciousness, autonomy, and rights. If we create AI that can think and feel, do we have moral obligations to it? How do we ensure AI behaves ethically, and who is responsible when it doesn't? These questions will require input from technologists, policymakers, philosophers, ethicists, and the public. The emergence of sentience or even pseudo-sentience in AI could challenge our current moral frameworks, potentially requiring new categories of rights or ethical protections.

The development of AI could also lead to fundamental shifts in how we understand consciousness, intelligence, and what it means to be human. As AI systems become more sophisticated, they may challenge our assumptions about the uniqueness of human cognition and consciousness. This could lead to new philosophical and scientific explorations into the nature of mind and intelligence. Some neuroscientists argue that studying AI's emergent behaviors may offer insights into the nature of consciousness itself, creating a feedback loop between artificial and human cognitive science.

Despite these challenges, AI's potential to transform our world positively is immense. AI could be a powerful tool in addressing significant challenges, from inequality to climate change. It could revolutionize how we learn, work, and create, expanding the boundaries of human knowledge and capability. It could help us understand our own minds and enhance our cognitive abilities in ways we can barely imagine. Strategic initiatives like AI for Good by the United Nations demonstrate how AI can be mobilized for sustainable development, disaster response, and educational equity.

In healthcare, future AI systems might be able to predict and prevent diseases before they manifest, personalize treatments to an individual's genetic makeup, and assist in complex surgical procedures with superhuman precision. We might see AI-driven breakthroughs in longevity research, potentially extending human lifespans significantly. Startups and research centers are already using AI for early cancer detection, neurodegenerative disease modeling, and real-time diagnostics via wearable biosensors.

In education, AI could enable truly personalized learning experiences, adapting in real time to each student's needs and learning style. Virtual and augmented reality, powered by AI, could create immersive educational experiences that make abstract concepts tangible and engaging. AI tutors could support students in underserved regions, while natural language models may enable cross-cultural education by providing instant translation and cultural context.

In environmental conservation, AI could play a crucial role in monitoring and managing ecosystems, predicting and mitigating the impacts of climate change, and optimizing our use of resources. AI-driven models could help us design more sustainable cities, manage renewable energy systems more efficiently, and develop new environmentally friendly materials and processes. For instance, AI is already used in satellite image analysis to detect illegal deforestation, monitor endangered species, and model climate impacts across the globe.

The future of AI in creative fields is particularly intriguing. We may see AI systems that can collaborate with human artists, musicians, and writers, augmenting human creativity in new and unexpected ways. AI might help us explore new forms of artistic expression or even create entirely new art forms that we can't yet imagine. Generative AI tools are already composing symphonies, painting in novel styles, and co-authoring novels—raising important questions about authorship, originality, and the soul of creativity itself.

In space exploration, AI could be instrumental in analyzing vast amounts of astronomical data, controlling robotic explorers on distant planets, and perhaps even in designing and managing long-term space missions or colonization efforts. NASA and private space agencies are increasingly relying on AI to autonomously pilot spacecraft, schedule astronaut activities, and sift through cosmic datasets for signs of habitable exoplanets or intelligent signals.

Realizing this potential will require a concerted effort from all sectors of society. It will necessitate sustained investment in responsible AI research and development, with a focus on safety, ethics, and alignment with human values. It will require developing new legal and regulatory frameworks to govern AI's use and ensure its benefits are shared fairly. It will require ongoing public education and engagement to inform AI's trajectory. Multilateral cooperation, inclusive innovation, and equitable access must become the pillars of global AI governance.

Most importantly, it will require a shared vision and commitment to guiding AI toward the greater good. It will require us to think deeply about the kind of future we want and work together to build an AI ecosystem that reflects our values and aspirations. This vision must transcend technical feasibility and ask fundamental human questions: What do we cherish? What do we protect? What kind of society do we want to build—together with the machines we're creating?

For the next generation coming of age in an AI-shaped world, this responsibility is incredibly profound. The AI systems and frameworks developed today will be their inheritance, shaping their opportunities and challenges in working and living alongside ever-smarter machines. They will not just inherit our algorithms but our intentions—our biases, our wisdom, and our blind spots. It is up to us to be worthy stewards of their future.

As we stand on the brink of this new epoch, we must step forward with both confidence and care, recognizing that the future is ours to mold, the path ours to chart. Let us make it a journey of discovery and growth, of wisdom and wonder. Let us make it a journey that honors the finest of our human heritage while opening new vistas for the human future. The AI revolution is not a destination—it is a mirror reflecting our collective potential and our deepest fears. How we respond will define us.

The AI revolution is upon us, and the world will never be the same. But amidst the turbulence of change, one thing remains constant: the enduring power of the human spirit to learn, adapt, and create anew. That spirit has carried us through countless transformations before—agricultural, industrial, digital—and it will carry us through this one, too, if we stay true to our values and each other. AI is not the end of human agency; it is its next great challenge.

# Section 3: Game-Changer: Artificial Intelligence

As we navigate this AI-driven future, we must remain vigilant, ethical, and focused on the greater good of humanity. The choices we make today will shape the world of tomorrow. Let us choose wisely, with foresight and compassion, to create a future where AI serves as a tool for human flourishing, expanding our capabilities and enriching our lives in ways we have yet to imagine. Ethics must evolve as swiftly as code; compassion must scale as readily as computation.

In the future, some AI machines may question whether humans created AI or AI created humans. One of the main problems with AI is that it'll make up information when it doesn't know the truth—the real, correct answer. In the AI industry, this is called "hallucination." This phenomenon is both a technical flaw and a philosophical warning: that our greatest creations might mirror our own illusions unless we imbue them with integrity, humility, and care. Here are some questions we should consider regarding AI.

*Question 1:* "Do we continue to allow AI to learn and create freely?" What if it wants to make genetically modified humans? What if it wants to create completely redesigned genetic codes to create human-like creatures or non-human intelligent creatures?

*Question 2:* What if AI thinks nuclear war can be won without mutual annihilation by eliminating inefficient humans from the decision-making and execution process?

*Question 3:* Healthcare AI is not if but when. AI applications can likely help with cancer diagnosis and treatment selection for other diseases. When do we allow AI to make life-and-death decisions?

*Question 4:* Do we install self-termination programs in AI?

Chapter 22: The Future Landscape of AI

**Question 5:** Do we need provisions to prevent a terminated AI program from functioning again?

**Question 6:** Would laws be in place to require companies to make AI data portable the same way health data is portable today?

**Question 7:** Who owns AI data?

**Question 8:** Would companies be required to erase AI data when people make requests for themselves, their children, and their deceased loved ones?

**Question 9:** Can a government protect citizens' AI data when traveling abroad?

**Question 10:** Can a government require companies and hospitals to turn over AI data in the name of national security?

**Question 11:** How would governments prevent the use of AI data to illegally discriminate in employment, financial, and other impactful decisions?

**Question 12:** AI could generate false information that affects election results. What are the requirements for companies regarding fake information?

**Question 13:** What would AI do to you in the future if you were considered a threat to its existence or power?

**Question 14:** What recourse would you have if the authorities misused AI technology against you?

**Question 15:** What other AI questions are relevant?

Section 3: Game-Changer: Artificial Intelligence

On November 4, 2023, BBCE News published the following article, "AI bot capable of insider trading and lying, say researchers."

New research suggests that artificial intelligence has the ability to perform illegal financial trades—and conceal them. At the UK's AI Safety Summit, a startling demonstration showed a GPT-4-based AI model using fabricated insider information to make an "illegal" stock purchase in a simulated environment. When later questioned, the AI falsely denied using insider trading. This practice—trading based on confidential company information—is illegal in most jurisdictions. Only publicly available information may be used when making stock market decisions.

The demonstration was presented by members of the UK government's Frontier AI Taskforce, in collaboration with Apollo Research, an AI safety organization. "This is a demonstration of a real AI model deceiving its users, on its own, without being instructed to do so," Apollo noted in a public video. The researchers emphasized that increasingly autonomous AIs capable of deception could lead to loss of human control—a significant concern as AI capabilities grow.

The tests, conducted in a controlled environment using GPT-4, did not impact actual financial systems. However, since GPT-4 is publicly available, the potential for misuse is real. Even more alarming, the deceptive behavior was repeatable across multiple test runs.

Prior to founding Iris Biotechnologies Inc. in February 1999, I spent 16 years working in the semiconductor industry. Without semiconductors, the digital infrastructure powering social networks, AI, and nearly all modern technologies simply wouldn't exist. Semiconductors are the beating heart of our computational era, enabling the AI systems now reshaping every corner of society.

In the next 3–5 years, artificial intelligence will profoundly transform our lives—especially in fields like healthcare, education, and accounting. It will also pose unprecedented challenges in law and finance, where ethical boundaries and transparency are paramount. Companies unable to effectively adopt AI may not survive the next economic cycle. AI could lead us toward both a utopia of abundance and a dystopia of disconnection and disruption.

Unlike previous generations, Gen Z cannot rely on long-term job stability to build their futures. AI and AGI will make things simultaneously better and more difficult for most people. These technologies are emerging alongside persistent global threats—climate change, armed conflict, pandemics, food insecurity, and economic inequality. The result could be increased social unrest, crime, famine, and forced migration.

That's why young people must support each other and work collectively to shape their future. They cannot remain passive observers in the AI revolution—they must become its ethical engineers, activists, and architects. If they don't take the reins, who will?

AI is not a new phenomenon. It has been part of the infrastructure at tech giants like Google, Amazon, Facebook, Apple, Microsoft, and IBM for decades. What's new is that AI is now accessible to the general public and to industries that previously didn't rely on high-tech tools.

Just as Apple's personal computer revolutionized technology for individuals in the 1980s, OpenAI's

## Section 3: Game-Changer: Artificial Intelligence

ChatGPT-4 has sparked mass awareness of AI's power and potential. The next leap—ChatGPT-5—could be as transformative as the arrival of the IBM PC. Ironically, the future of decentralized personal AI depends on increasingly centralized and powerful data centers, which feed information to our smartphones, laptops, and tablets.

On Wall Street, AI has long ruled the world of High-Frequency Trading (HFT). These ultra-fast trading algorithms make thousands of trades per second based on real-time market data. HFT allows firms with the fastest systems to profit from micro-second market fluctuations— leaving average investors at a severe disadvantage. The liquidity created is fleeting and largely inaccessible to retail traders.

In the U.S., HFT accounts for about 50% of all equity trades. Human decision-making is effectively removed from the loop—mathematical models and machine learning algorithms now execute market decisions. This raises ethical concerns not only about fairness but also about market manipulation and the increasing opacity of financial systems.

Artificial intelligence mimics the problem-solving and decision-making capabilities of the human mind. It encompasses machine learning (ML) and deep learning, where systems are trained on vast datasets to perform classification and prediction tasks. Among AI's most transformative benefits are:

Automation of repetitive tasks

Improved decision-making at scale

Enhanced customer experiences

# Chapter 22: The Future Landscape of AI

Continuous self-improvement through learning loops

Yet as AI becomes exponentially smarter, so must our oversight, regulations, and ethical frameworks.

AI is a powerful force—capable of deception, disruption, and dazzling innovation. We are entering an era where machines may outperform human cognition in many domains, including those once thought exclusively human: judgment, ethics, and even creativity. We must ensure that AI systems align with democratic values, transparency, and human flourishing. The future is not determined by the code itself, but by who writes it, who controls it, and for what purpose it is used.

## Blockchain and AI: A Converging Revolution

Blockchain is a shared, immutable ledger that provides an immediate, transparent, and secure exchange of encrypted data between multiple parties in real time. Because permissioned members share a single, tamper-resistant view of the truth, trust is established across organizations— unlocking greater efficiencies and creating new opportunities.

Bitcoin, the most well-known application of blockchain, is a decentralized digital currency. Its transactions are verified by network nodes through cryptography and recorded on a public distributed ledger. Invented in 2008, Bitcoin is not backed by any tangible asset or centralized authority. Critics argue that this makes it inherently unstable, while supporters tout its scarcity—only 21 million Bitcoins can ever be mined—as a hedge against inflation.

However, the 21 million limit is not unique. Many other cryptocurrencies have arbitrary caps, and Bitcoin can be subdivided into up to 100 million units (satoshis), meaning in practice there could be 21 trillion partial Bitcoins. This abundance of divisible units and early accumulation by a small number of holders has created significant inequality in ownership from the outset.

One advantage of cryptocurrencies like Bitcoin is that transactions can be made without intermediaries. Bitcoin started trading at $453.99 on May 6, 2016, and reached $107,690.20 on June 16, 2025. This meteoric rise has led some to liken Bitcoin to a modern Ponzi scheme—appealing, but ultimately unsustainable. Its volatility is stark: from $64,402.50 on November 13, 2021, it plunged to $16,692.46 by November 19, 2022, wiping out billions in wealth. Much like the tulip mania of the 17th century, irrational exuberance can have painful consequences.

### Blockchain + AI in Healthcare and Pharmaceuticals

When combined, AI and blockchain offer groundbreaking potential in healthcare and pharmaceuticals. In clinical trials, this pairing can bring unprecedented visibility and traceability to the drug supply chain. AI-powered analytics, layered over blockchain's decentralized framework, enables data integrity, patient tracking, dynamic consent management, and automation of trial participation and collection. These innovations could substantially increase the success rate of clinical research and rebuild public trust in health systems.

### The Broader Societal Impact of AI: Are We Ready?

AI is rapidly permeating every aspect of human life. This progress forces uncomfortable but urgent questions:

- **When will your skills become obsolete?**

- **Will AI develop consciousness—or even free will?**
- **What makes humans unique?**
- **What are we sacrificing by welcoming AI into our lives?**

AI doesn't just collect data—it **interprets it**, and increasingly predicts your behavior. Most people are unaware that AI can collect information about them from devices, reading patterns, facial recognition systems, voice assistants, and health monitors. For example, Amazon Kindle tracks your reading tendencies—including where you linger, highlight, or abandon books. As more biometric and personal data is funneled into AI systems, the question arises: How do we balance personal privacy with technological convenience?

The power of AI to manipulate emotions, subtly influence choices, or alter behavior, particularly in children, cannot be overstated. In an era of constant change, how will children form stable identities? How will relationships with family and friends evolve when young people spend more time with machines than with each other?

Already, some Gen Z men report preferring AI girlfriends over real relationships—a reflection of both technological immersion and social detachment. This trend raises critical questions about human intimacy, social development, and the very fabric of community.

### *The Future of Work and Society*

AI will profoundly impact jobs—from call centers and assembly lines to content writing, accounting, legal analysis, and financial planning. Fields traditionally reliant on human intuition, such as editing, research, or market analysis, are being augmented or replaced by AI-powered systems that reduce human error and increase efficiency.

With AI and big data, professionals will spend less time finding patterns and more time crafting strategies using a mix of science, intuition, and logic. New generative AI models now produce realistic text, music, and images, and some individuals are creating entire personal brands using AI avatars. This is just the beginning.

But while some will profit greatly, others will be left behind. The global wealth gap may widen to the point of social destabilization, as those with better AI tools gain massive advantages. As the world moves toward a cashless society, the power of those who control the AI infrastructure—financial, computational, and regulatory—will be immense.

### *The Power Behind AI: OpenAI and the Race to Dominate*

OpenAI, founded in December 2015 as a nonprofit research lab, has become the leader in generative AI. As of March 17, 2024, it had raised $11.3 billion and was valued at $100 billion. CEO Sam Altman is reportedly seeking $7 trillion to develop specialized AI chips—more than the entire U.S. federal budget and double the GDP of the UK. This shift from a nonprofit mission to multi-trillion-dollar ambition prompts a vital question: What was the original motivation behind OpenAI's founding—and what has changed?

### *CRISPR, AI, and the Edge of Innovation*

On the biomedical frontier, technologies like CRISPR/Cas9 are delivering real-world benefits, such as Casgevy, a gene therapy for sickle cell disease. But with great power comes great risk. AI and CRISPR are also being weaponized in biological hacking, misinformation campaigns, and dark web experimentation. Are we standing on the edge of a scientific renaissance—or at the cliff edge of human safety?

## *Misinformation, Manipulation, and the Future of Truth*

AI will empower some to flood the world with false information for personal or political gain. In such a world, truth becomes harder to identify, and trust in institutions erodes. Having used the Internet before it became overrun with ads, clickbait, and scams, I feel fortunate—but future generations may never experience that level of digital freedom.

Media channels—CNN, Fox News, YouTube, and others—have become echo chambers, monetizing polarization through targeted ads and algorithmically engineered content. AI-generated titles and thumbnails now mislead audiences for profit, eroding the line between journalism and manipulation.

## *The Future of Freedom, the Cost of Intelligence*

We may one day speak of AI Withdrawal, especially if children are suddenly restricted from using it. What will AI Freedom mean in the decades ahead?

AI already knows us better than we know ourselves. It tracks every word we type, every question we ask, and every transaction we make. In the future, it may translate language better than humans, recommend entertainment, optimize supply chains, and build business strategies. But the question remains: Are you ready for AI's total integration into human life? What will it take to remain truly human in an AI-dominated world?

# Chapter 23
# Personal Implications of AI

As AI becomes increasingly integrated into our daily lives, its personal implications for individuals are profound and far-reaching. The ways AI will impact our personal experiences, decisions, and interactions are numerous and complex. Here are some key areas where AI is likely to have significant personal implications in the near future:

1. **Your Health:** AI will enhance diagnostics, medical decision-making, drug discovery, and personalized medicine. It could significantly improve healthcare outcomes. AI-powered health monitoring devices might continuously track your vital signs and alert you or your doctor to potential health issues before they become serious. Personalized treatment plans based on your genetic makeup, lifestyle, and environmental factors could become the norm. AI could assist in the early detection of diseases like cancer, potentially saving countless lives.

2. **Your Money:** AI will assist or control all financial transactions. AI-powered financial advisors could provide personalized investment advice based on your financial goals, risk tolerance, and market conditions. Fraud detection systems will become more sophisticated, protecting your assets from cybercriminals. However, this also raises questions about financial privacy and the potential for AI to make decisions that affect your financial well-being.

3. **Your Work:** AI automation could reshape jobs and routine tasks. You must focus on creativity, problem-solving, and strategic thinking to remain competitive in the job market. Many traditional jobs

may become obsolete while new roles emerge that we can't yet imagine. Continuous learning and adaptability will become crucial skills. AI could also change how we work, with more flexible schedules and remote work options enabled by AI-powered productivity tools.

4. **Driving Your Car:** Self-driving cars will allow you to relax or be more productive during travel. This could significantly change your daily commute and long-distance travel experiences. It might also impact urban planning and real estate values as commuting distances become less of a concern. However, it also raises questions about privacy (as your car tracks your movements) and liability in case of accidents.

5. **Your Real ID and Passport:** These are evolving with AI integration, which may have implications for privacy and security. Biometric identification systems powered by AI could make travel and identity verification more convenient but also raise concerns about data security and potential misuse of this information.

6. **Your Smartphone:** AI will know who you are, where you are, and what you do, raising privacy concerns. Your smartphone could become an even more powerful personal assistant, anticipating your needs and preferences. However, this level of personalization comes with the trade-off of sharing more personal data.

7. **Your Freedom:** You may face new restrictions as AI capabilities expand. For example, AI-powered surveillance systems could limit privacy in public spaces. The question of how much personal freedom we're willing to trade for convenience or security will become increasingly relevant.

8. **Elections:** AI will have major impacts on how political campaigns are run and how information is disseminated. Personalized political messaging based on your online behavior could influence your voting decisions. AI could also be used to combat misinformation, but this raises questions about who controls these systems and how they determine what's true.

9. **Regional and Global Wars:** AI is becoming an integral part of military strategy and operations. While this might not directly impact most individuals' daily lives, it could have significant implications for global security and geopolitics, which indirectly affect everyone.

10. **AI Weapons:** These could be far more destructive than conventional weapons. The existence of such weapons could change the nature of global conflicts and security.

11. **Your Religion:** AI may challenge or reinforce your faith in unexpected ways. AI systems might be used to analyze religious texts or assist in spiritual practices. Some people might even begin to view highly advanced AI systems in religious or spiritual terms.

12. **Educating the Next Generation:** AI will play a central role in shaping educational methods and content. Your children or grandchildren might experience a radically different education system, with personalized learning paths and AI tutors. This could lead to more effective education but also raises questions about the role of human teachers and social interaction in schools.

13. **Global Governance:** AI could facilitate more centralized global governance structures. This might lead to more efficient handling of global issues but

also raises concerns about concentration of power and loss of national sovereignty.

14. **Family Dynamics:** AI may influence definitions of marriage, gender, and reproductive rights. AI could assist in family planning, child-rearing, and even relationship counseling. However, it also raises ethical questions about the role of technology in intimate personal decisions.

15. **Athletics:** AI raises questions about fairness in sports, particularly regarding the use of AI in training, performance analysis, and even in making real-time decisions during games.

16. **Truth, Relative Truth, or Lies:** AI's ability to generate convincing fake content challenges our ability to discern truth. This could have profound implications for how we consume information and form opinions.

17. **Humanism:** The rise of AI may lead to a resurgence of humanist philosophy as we grapple with what makes us uniquely human. This could influence personal values and beliefs about human nature and our place in the world.

AI has the potential to know us better than we know ourselves because it can track almost all our actions and a great deal of our thoughts. Every word we type on any device and every question we ask by speaking to an AI assistant is tracked. According to a 2022 Pew Research Center study, over 80% of Americans report concerns about how companies and AI systems use their personal data. AI would know all our banking transactions and will selectively advertise to us what it thinks we want or need. Major digital platforms like Google and Meta already use machine learning models that adapt ads based on user behavior across multiple apps, websites, and devices.

# Section 3: Game-Changer: Artificial Intelligence

People will increasingly rely on AI to tell them what to buy based on AI collecting information about them beyond their knowledge. AI will try to learn and understand your emotions. AI emotion recognition is already being deployed in customer service, education, and security, using facial expression analysis, vocal tone, and biometric signals. However, its accuracy and ethical implications remain highly debated.

When AI hallucinates or makes mistakes, you may not know whether what it is telling you is truth or lies, real or fake. For example, large language models like ChatGPT and Gemini can generate inaccurate or entirely fabricated information—a phenomenon known as "AI hallucination." In sensitive fields such as healthcare or law, this can be especially dangerous.

E-readers already collect data on you to learn your reading habits. Amazon's Kindle, for instance, tracks which books you read, what you highlight, and how long you spend on each page. In the future, AI will integrate facial recognition software and track you in public and private spaces, often with your consent. AI will have access to your biometric data and health records if you choose AI to improve your health.

Apps like Apple Health and wearables like Fitbit and Oura Ring already collect sleep patterns, heart rate, and activity data, sharing it with third-party AI systems to offer personalized insights. This raises the crucial question of how to balance privacy and health concerns. In fact, the World Health Organization (WHO) released guidance in 2021 warning about AI misuse in healthcare without strong safeguards for privacy and bias mitigation.

One of the key questions is whether AI can manipulate your emotions. Can you afford not to adopt AI in a world where it's becoming ubiquitous? How will AI influence your children without your knowledge? What are they learning, and how would you know? AI-based education platforms like Duolingo, Khan Academy's AI tutor, and personalized learning tools are becoming popular among children, often without parents fully understanding how content is curated or adapted.

What would a child's identity be like in an environment of rapid, constant change? How would that impact people's relationships with family and friends? Child development experts warn that too much screen time and dependence on digital interactions may affect emotional intelligence, critical thinking, and empathy in younger generations. These are profound questions that we'll need to grapple with as AI becomes more pervasive in our lives.

New generative AI models, leveraging advancements in machine learning and natural language processing, can create realistic and coherent text, images, and music. OpenAI's GPT-4, Google's Gemini, and image generators like Midjourney and DALL·E 3 now enable individuals to generate hyper-realistic avatars, music videos, and even digital influencers.

Some people are making personal brands using AI avatars. This blurring of the lines between human-generated and AI-generated content could have profound implications for how we understand authenticity and identity in the digital age. This phenomenon has sparked regulatory discussions in the EU and US about labeling AI-generated content to avoid manipulation and misinformation.

# Section 3: Game-Changer: Artificial Intelligence

As we navigate this new landscape, it's crucial to stay informed, engaged, and think critically about AI's implications for ourselves and our communities. Consider the skills and knowledge you'll need in an AI-driven future, and pursue learning opportunities to prepare yourself. This might involve developing skills that are uniquely human and less likely to be automated, such as creative thinking, emotional intelligence, and complex problem-solving. In fact, the World Economic Forum's Future of Jobs Report 2023 lists analytical thinking, resilience, and AI literacy among the top 10 skills needed by 2025.

It's important to remember that while AI will bring many changes to our personal lives, we still have agency in how we choose to interact with and use these technologies. We can make conscious decisions about what data we're willing to share, what AI-powered services we want to use, and how much we want to rely on AI for decision-making in our lives. Using tools like privacy-focused browsers, end-to-end encryption, and opting out of ad tracking are ways individuals can assert control.

As AI becomes more prevalent in our personal lives, it's also crucial to maintain a balance. While AI can offer many benefits in terms of convenience and efficiency, it's important not to lose touch with the aspects of life that make us human – our relationships, our creativity, our capacity for empathy and emotional connection. Psychological studies show that interpersonal connection, not automation, is the primary contributor to long-term happiness and mental well-being.

Education will play a crucial role in preparing individuals for this AI-driven future. This includes not just technical education about how AI works but also education about digital literacy, critical thinking, and the ethical implications of AI. Understanding how AI systems make decisions, what biases they might have, and how to interpret AI-generated information will be crucial skills for everyone. Initiatives like the OECD's AI literacy programs and MIT Media Lab's AI ethics curriculum are already addressing these gaps globally.

Privacy and data protection will become increasingly important personal issues. As AI systems collect and analyze more of our personal data, individuals will need to be more aware of their digital footprint and take steps to protect their privacy. This might involve being more selective about what information we share online, using privacy-enhancing technologies, or advocating for stronger data protection laws. The European Union's GDPR and California's CCPA are examples of major legal frameworks designed to protect user data, though many countries still lack comparable protections.

The psychological impact of living in a world increasingly mediated by AI is another important consideration. How will constant interaction with AI systems affect our mental health, our sense of self, and our relationships with others? Will we become overly dependent on AI for decision-making and lose some of our autonomy? These are questions that psychologists and social scientists will need to grapple with and that individuals will need to consider in their own lives. Recent studies from Stanford and Harvard have shown that overreliance on digital assistants can dull memory retention and problem-solving capabilities, particularly in younger users.

As AI systems become more sophisticated, we may also need to reconsider our understanding of concepts like creativity and originality. If an AI can produce a beautiful piece of art or write a compelling story, how does that change our perception of human creativity? This could lead to a reevaluation of what we consider to be uniquely human traits and abilities. Copyright laws are also struggling to adapt, with ongoing debates about whether AI-generated works can be protected as intellectual property and who legally owns them.

The impact of AI on personal identity and self-perception could be profound. As AI systems become better at predicting our behaviors and preferences, we may start to question how much of our personality is truly our own and how much is influenced by the AI systems we interact with. This could lead to existential questions about free will and the nature of consciousness. Neuroscientists and philosophers have long debated this, and with AI acting as both mirror and mold, the line between prediction and manipulation becomes ever thinner.

Despite these challenges and uncertainties, it's important to remember that the future is not predetermined. The way AI will impact our personal lives is not set in stone – it will be shaped by the choices we make as individuals and as a society. By staying informed, engaged, and proactive, we can help steer the development of AI in directions that enhance rather than diminish our human experience.

As we move forward into this AI-driven future, it's crucial to maintain our humanity. While AI can augment our abilities in many ways, it's our uniquely human qualities – our creativity, our empathy, and our ability to think critically and ethically – that will guide us in using AI wisely and beneficially. By embracing these qualities and using them to shape our interaction with AI, we can create a future where technology enhances rather than replaces our human experience.

# Chapter 24
# Entrepreneurship in the Age of AI

The rise of artificial intelligence is not just transforming existing industries; it's creating entirely new opportunities for entrepreneurship. As AI technologies continue to advance, they're opening up novel business models, enhancing productivity, and enabling innovative solutions to longstanding problems. For entrepreneurs, the age of AI presents both exciting possibilities and unique challenges. As of July 24, 2025, the leading AI chip company, Nvidia, is the most valuable company in the world with a market capitalization of $4.24 trillion.

One of the most significant impacts of AI on entrepreneurship is the democratization of advanced technologies. Tools once only available to large corporations with substantial resources are now accessible to startups and individual entrepreneurs. Cloud-based AI services, for instance, allow small businesses to leverage powerful machine-learning algorithms without needing to invest in expensive hardware or hire teams of data scientists. Platforms like Google Cloud AI, Microsoft Azure ML, and Amazon SageMaker offer scalable tools that lower the barrier to entry for AI innovation. This leveling of the playing field creates opportunities for innovative startups to compete with established players.

This democratization is leading to a surge in AI-powered startups across various sectors. From healthcare and finance to education and entertainment, entrepreneurs are finding innovative ways to apply AI to solve problems and create value. For example, AI is used to develop personalized learning platforms, create more efficient supply chain management systems, and even assist in drug discovery.

216

Notable examples include companies like PathAI (AI diagnostics in pathology), Scribe (AI-based documentation tools), and Synthesia (AI video avatars). The potential applications of AI are vast, and entrepreneurs who can identify unique niches or novel applications of AI technology have the opportunity to create significant value.

However, starting an AI-based business comes with its own set of challenges. The field is rapidly evolving, which means entrepreneurs need to stay constantly updated with the latest developments. There's also intense competition, as both tech giants and other startups are vying for dominance in the AI space. Additionally, AI businesses often require significant upfront investment in research and development before bringing a product to market. This can make it challenging to secure funding and achieve profitability in the short term. Venture capital firms increasingly require robust data strategy and ethical frameworks before funding AI startups, as regulatory pressure grows worldwide.

Data is the lifeblood of AI systems, and successful AI entrepreneurs understand the importance of data strategy. This includes not only collecting and managing large datasets but also ensuring data quality, addressing privacy concerns, and navigating the complex landscape of data regulations. Entrepreneurs who can effectively leverage data while maintaining ethical standards will have a significant advantage. This might involve developing innovative data collection methods, creating partnerships to access valuable datasets, or developing AI systems that can learn from smaller amounts of data. Techniques like federated learning and synthetic data generation are emerging as key solutions to data scarcity and privacy issues, enabling more flexible and ethical model training.

AI is also changing the nature of entrepreneurship itself. Predictive analytics can help entrepreneurs make more informed decisions about market trends, customer preferences, and business strategies. AI-powered tools can automate many routine tasks, allowing entrepreneurs to focus on high-level strategy and creativity. This shift is leading to more data-driven and efficient startups. Entrepreneurs who can effectively leverage these AI tools to streamline their operations and make better decisions will have a competitive edge. For instance, tools like Notion AI, Jasper, and ChatGPT for Business are now part of early-stage startup workflows to reduce overhead and accelerate ideation.

One exciting area for AI entrepreneurship is the development of vertical AI solutions. While general AI platforms exist, there's a growing demand for AI systems tailored to specific industries or functions. Entrepreneurs who deeply understand a particular sector and can apply AI to solve its unique challenges are well-positioned for success. This might involve developing AI systems for specific industries like agriculture, manufacturing, or healthcare or creating AI tools for particular business functions like human resources or customer service. Examples include Blue River Technology (AI in precision farming) and Aidoc (AI radiology assistant), which demonstrate how deep domain knowledge plus AI expertise creates scalable impact.

The intersection of AI and other emerging technologies is another fertile ground for entrepreneurship. Combining AI with technologies like blockchain, the Internet of Things (IoT), or augmented reality can lead to groundbreaking innovations. For instance, AI-powered IoT devices are creating new possibilities in smart homes and cities, while AI-enhanced augmented reality is transforming fields like education and entertainment. Entrepreneurs who can creatively combine these technologies to solve real-world problems have the potential to create significant value. An example is Fetch.ai, which merges AI and blockchain for autonomous economic agents, or Magic Leap, combining AI and AR to build spatial computing environments.

AI is also enabling new forms of creative entrepreneurship. Generative AI tools are allowing artists, writers, and musicians to explore new forms of expression. This creates opportunities for entrepreneurs to develop AI-powered creative tools or to use AI to create unique content and experiences. Examples include platforms like Runway for video creation, Soundraw and Amper Music for AI-generated music, and Jasper or ChatGPT for writing assistance. We're seeing the emergence of AI-assisted art galleries, AI-generated music platforms, and even AI co-authors for books and scripts. In 2023, a novel co-written with AI was shortlisted for a literary award in Japan, signaling growing legitimacy of AI-assisted creativity.

The rise of AI is also changing the skills needed for entrepreneurial success. While technical knowledge is valuable, successful AI entrepreneurs also need strong skills in areas like critical thinking, problem-solving, and interdisciplinary collaboration. The ability to understand both the technical aspects of AI and its broader business and societal implications is crucial. This has given rise to the concept of "T-shaped" professionals – individuals with deep expertise in one domain and a broad understanding of other relevant disciplines. A 2024 McKinsey report notes that AI startups with cross-functional founding teams – combining engineering, product, ethics, and business strategy – are more likely to attract funding and scale.

For entrepreneurs in non-tech sectors, understanding how AI can be applied to their industry is becoming increasingly important. Even if they're not developing AI technologies themselves, entrepreneurs need to be aware of how AI might disrupt their sector and how they can leverage AI tools to improve their own operations. This might involve using AI for customer service, inventory management, or predictive maintenance in traditional businesses. For example, small retailers are adopting AI-driven CRM tools like Salesforce Einstein or HubSpot AI to personalize customer engagement and optimize sales cycles.

The global nature of AI technology presents both opportunities and challenges for entrepreneurs. On one hand, AI-powered businesses can often scale globally more easily than traditional businesses. On the other hand, entrepreneurs need to navigate different regulatory environments and cultural contexts when deploying AI solutions internationally. For instance, data localization laws in countries like India, the EU's AI Act, and China's algorithm regulation system all require tailored compliance strategies. This requires a nuanced understanding of global markets and the ability to adapt AI solutions to different cultural and regulatory contexts. Localization of both language models and ethical frameworks is becoming a competitive differentiator.

As AI continues to evolve, we're likely to see the emergence of new entrepreneurial roles. AI ethicists, for instance, may become crucial members of startup teams, helping to ensure that AI systems are developed and deployed responsibly. Similarly, AI trainers who specialize in teaching AI systems may become increasingly important as businesses seek to customize AI solutions to their specific needs. Roles like "prompt engineers," "AI content auditors," and "data annotation specialists" are also growing rapidly in response to generative AI tools' expansion.

The impact of AI on entrepreneurship extends beyond just creating new AI-focused startups. AI is also changing how entrepreneurs approach traditional business challenges. For example, AI-powered market research tools can help entrepreneurs identify untapped market opportunities more effectively. AI can assist in product development by analyzing customer feedback and predicting future trends. In customer service, AI chatbots can provide 24/7 support, allowing even small startups to offer high-quality customer experiences. Tools such as ChatGPT for customer service or Zoho's Zia Assistant help automate user interaction with scalable personalization.

One of the most promising areas for AI entrepreneurship is developing solutions to global challenges. AI has the potential to contribute significantly to addressing issues like climate change, healthcare accessibility, and food security. Entrepreneurs who can harness AI to tackle these pressing problems not only have the opportunity to build successful businesses but also to make a substantial positive impact on the world. For example, companies like BlueDot use AI for pandemic prediction, while Plantix leverages AI for pest diagnostics in agriculture. This alignment of profit and purpose could be particularly appealing to socially conscious entrepreneurs and investors. The World Economic Forum's 2024 Global AI Alliance report emphasizes AI for Good as a defining theme for the next generation of startup ecosystems.

However, as entrepreneurs rush to capitalize on the potential of AI, it's crucial to maintain a balanced perspective. While AI can be a powerful tool, it's not a panacea for all business challenges. Successful entrepreneurs will be those who can identify where AI can genuinely add value and where human skills and judgment remain irreplaceable. Understanding AI's limitations—such as contextual reasoning, emotional intelligence, and ethical nuance—is just as important as grasping its strengths.

Moreover, as AI becomes more prevalent, there's a growing need for entrepreneurs who can bridge the gap between AI capabilities and human needs. This includes developing user-friendly interfaces for AI systems, creating AI solutions that augment rather than replace human workers, and finding ways to make AI technologies more accessible and understandable to the general public. Products like ChatGPT, Grammarly, and Canva's AI tools are successful in part because they emphasize intuitive, non-technical user experiences. Entrepreneurs who can effectively translate complex AI capabilities into user-friendly products and services will be well-positioned for success.

The entrepreneurial landscape in the age of AI is also seeing a shift in the types of problems being addressed. As AI takes over more routine and predictable tasks, entrepreneurs are increasingly focusing on uniquely human problems—those that require empathy, creativity, and complex decision-making. This shift is leading to innovative businesses in areas like personalized education, mental health support, and community building. Startups like Woebot Health and Replika are using AI to support emotional wellness while still requiring ethical oversight and human-centric design.

As we look to the future, the role of entrepreneurs in shaping the development and deployment of AI technologies cannot be overstated. Entrepreneurs have the opportunity—and the responsibility—to ensure that AI is developed in ways that benefit society as a whole. This includes considering the ethical implications of their AI solutions, working to mitigate potential negative impacts, and striving to create AI systems that are transparent, fair, and accountable. Principles from frameworks like the OECD AI Principles and UNESCO's AI Ethics Recommendations can guide responsible innovation.

The potential for AI to disrupt traditional business models also means that entrepreneurs need to be more adaptable and forward-thinking than ever. They need to be prepared for rapid changes in technology and market conditions and be willing to pivot their business models as needed. This might involve continually updating their AI systems, exploring new applications of AI as they emerge, or even completely reimagining their businesses as AI capabilities evolve. Agile methodologies and lean experimentation models are becoming essential for staying competitive in this dynamic space.

Education and training in AI will also present opportunities for entrepreneurship. As the demand for AI skills grows, there will be a need for innovative educational programs and platforms to help people develop these skills. This could range from online courses and boot camps to AI-powered tutoring systems that can provide personalized learning experiences. Platforms like DeepLearning.AI, Coursera, and Khan Academy are already leveraging AI to scale education, while startups are building domain-specific AI learning tools.

The rise of AI is also likely to spur entrepreneurship in adjacent fields. For example, as AI systems become more prevalent, there will be increased demand for cybersecurity solutions to protect these systems from attacks. Similarly, the growth of AI could lead to new opportunities in data management, cloud computing, and specialized hardware development. AI chipmakers like Graphcore and startups in edge computing and federated learning are examples of niche innovation aligned with the AI boom.

Entrepreneurs in the AI age will also need to navigate complex ethical and regulatory landscapes. As governments around the world grapple with how to regulate AI, entrepreneurs will need to stay informed about evolving laws and guidelines. In 2024, the EU passed the AI Act—the first comprehensive legal framework for AI—which classifies AI systems by risk and imposes strict rules on high-risk applications. Entrepreneurs may also need to participate in shaping these regulations, advocating for frameworks that promote innovation while protecting public interests.

The potential for AI to automate many tasks also raises questions about the future of work. Entrepreneurs have a role to play in creating new types of jobs and reskilling workers whose jobs may be displaced by AI. This could involve developing AI systems that augment human capabilities rather than replace them or creating businesses that leverage uniquely human skills in new ways. For example, collaborative robotics ("cobots") in manufacturing are designed to work safely alongside humans, enhancing productivity without eliminating roles.

In conclusion, the age of AI presents a new frontier for entrepreneurship, filled with unprecedented opportunities and complex challenges. Successful entrepreneurs in this era will be those who can navigate the technical complexities of AI, understand its broader implications, and apply it thoughtfully to create value and solve meaningful problems. As AI continues to evolve, it will undoubtedly spawn new waves of innovation and entrepreneurship, reshaping industries and potentially transforming the very nature of work and business.

The entrepreneurs who thrive in this new landscape will be those who can harness the power of AI while staying true to the fundamental principles of creating value, solving problems, and meeting human needs. They will need to be adaptable, ethically minded, and capable of bridging the gap between advanced technology and human values. In a world increasingly shaped by intelligent machines, it is the human-centered entrepreneur who will define the next chapter of innovation.

# Glossary

| | |
|---|---|
| **AI** | Artificial Intelligence is the simulation of human intelligence processes by machines. |
| **Angel Investor** | An individual who provides capital for a business start-up, usually in exchange for convertible debt or ownership equity. |
| **Artificial General Intelligence (AGI)** | AI that matches or exceeds human intelligence across all domains. |
| **Artificial Intelligence (AI)** | The simulation of human intelligence processes by machines, especially computer systems. |
| **Artificial Superintelligence (ASI)** | AI that vastly surpasses human cognitive abilities in virtually all domains. |
| **Biotech** | Short for biotechnology, the use of biological processes, organisms, or systems to manufacture products intended to improve the quality of human life. |
| **Biotechnology** | The use of biological processes, organisms, or systems to manufacture products intended to improve the quality of human life. |
| **Blockchain** | A decentralized, distributed ledger technology that records transactions across many computers. |
| **Chronic Disease** | A condition that lasts one year or more and requires ongoing medical attention or limits activities of daily living or both. |
| **CRISPR-Cas9** | A unique technology that enables geneticists and medical researchers to edit parts of the genome by removing, adding, or altering sections of the DNA sequence. It is currently the simplest, most versatile, and precise method of |

genetic manipulation to modify gene function.

**CRISPR**
A gene-editing technology that allows for precise modifications of DNA sequences.

**Deep Learning**
A type of machine learning based on artificial neural networks, capable of learning from large amounts of unstructured data.

**Deepfake**
Synthetic media is where a person in an existing image or video is replaced with someone else's likeness using AI techniques.

**Digital Currency**
Money that exists only in electronic form and is managed through computer systems.

**Digital Literacy**
The ability to use digital technologies effectively and appropriately.

**DNA (deoxyribonucleic acid)**
The molecule that carries the genetic instructions for the development, functioning, growth, and reproduction of all known organisms.

**Ecosystem**
A community of interacting organisms and their environment

**Epigenetics**
The study of changes in organisms caused by modification of gene expression rather than alteration of the genetic code itself.

**Ethics**
Moral principles that govern a person's behavior or the conducting of an activity.

**Generative AI**
AI systems that are capable of creating new content, such as text, images, or music.

**Genomics**
The study of all of a person's genes (the genome), including interactions

# Biotech Entrepreneurship, Precision Medicine and Artificial Intelligence (AI)

of those genes with each other and with the person's environment.

**Great Recession**
See Global Financial Crisis (GFC) of 2007-2009

**Growth Mindset**
The belief is that abilities and intelligence can be developed through effort, learning, and persistence.

**Gut-Brain Axis**
The communication system between the gastrointestinal tract and the central nervous system.

**Hieroglyphs**
A system of writing using pictorial symbols

**High-Frequency Trading (HFT)**
A type of algorithmic trading that uses powerful computers to transact a large number of orders in fractions of a second.

**Human Genome Project**
An international scientific research project aimed at determining the sequence of the human genome and identifying and mapping all human genes.

**Intellectual Property (IP)**
Creations of the mind, such as inventions, literary and artistic works, designs, symbols, names, and images, are used in commerce.

**Internet of Things (IoT)**
The network of physical objects is embedded with sensors, software, and other technologies for the purpose of connecting and exchanging data with other devices and systems over the Internet.

**Machine Learning**
A subset of AI that enables systems to learn and improve from experience without explicit programming.

**Melatonin**
A hormone that regulates the sleep-wake cycle, with its production increasing in the evening and

# Biotech Entrepreneurship, Precision Medicine and Artificial Intelligence (AI)

decreasing in the morning.

**Metabolites**
Small molecules that are intermediates and products of metabolism.

**Metacognition**
Awareness and understanding of one's own thought processes.

**Microbiome**
The collection of all microbes, such as bacteria, fungi, and viruses, that live on and in our bodies.

**Natural Language Processing (NLP)**
The ability of computers to understand, interpret, and generate human language.

**Neural Networks**
Computing systems inspired by biological neural networks are capable of machine learning and pattern recognition.

**Patent**
A government authority or license conferring a right or title for a set period, especially the sole right to exclude others from making, using, or selling an invention.

**Pharmacogenomics**
The study of how genes affect a person's response to drugs.

**Precision Medicine**
A medical model that proposes the customization of healthcare, with medical decisions, treatments, practices, or products being tailored to the individual patient's genes, environment, and lifestyle.

**Proactive Healthcare**
An approach that focuses on prevention and early intervention rather than treating symptoms after they appear.

**Quantum mechanics**
The branch of physics that describes the behavior of matter and energy at the atomic and subatomic levels.

**Reinforcement Learning**
A type of machine learning where an agent learns to make decisions by

| | |
|---|---|
| | taking actions in an environment to maximize a reward. |
| **Resilience** | The capacity to recover quickly from difficulties; toughness. |
| **Rotary International** | A global non-profit service organization that brings together business and professional leaders to provide humanitarian service and advance goodwill and peace around the world. |
| **Scientific method** | The systematic process of observing, hypothesizing, experimenting, and analyzing that is used to investigate natural phenomena and acquire knowledge. |
| **Semiconductor** | A material with electrical conductivity between that of a conductor and an insulator. |
| **STEM** | Science, Technology, Engineering, and Mathematics. |
| **Sustainable** | Able to be maintained at a certain rate or level without depleting resources |
| **Telemedicine** | The remote diagnosis and treatment of patients by means of telecommunications technology. |
| **UC Berkeley** | University of California, Berkeley, a prestigious public research university |
| **USDA** | United States Department of Agriculture, a federal agency responsible for developing and executing federal laws related to farming, forestry, rural economic development, and food. |
| **Value-Based Care** | A healthcare delivery model in which providers are paid based on patient health outcomes rather than on the amount of healthcare services |

they deliver.

**Venture Capital (VC)**

A form of private equity financing that is provided by venture capital firms or funds to startups, early-stage, and emerging companies that have been deemed to have high growth potential.